Captain Mark Styler's, The Masquerades

Book one

Robert David Jensen

Table of Contents

List of Characters

Captain Mark Styler: Pilot, electrical engineer, owner of the *Marvel,* a merchant ship.

Chantal: Android, pilot, doctor.

Sara: The ship's computer.

Eric Styler: Mark's father.

Nana: Android, Mark's nanny.

Corrie Aileen Styler: Mark's mother, Eric's second wife.

Melissa: Meloson Alien.

Walburgis: Alien.

Lord Jayson Stevenson: Prince, Mark's former employee, and brother to Katherine.

Katherine Stevenson: Princess, part owner of the *Marvel,* and wife to Mark.

Crystal Styler: Daughter of Mark and Katherine.

Bill: Cook, labour, cargo specialist.

Mason McDonnell: Drive engineer specialist.

Timothy Andorra: Mark's former employee.

Tim: Timothy's son, an electrical technical apprentice.

Misty: Genetically enhanced female dog.

Andy: Mark's former android employee.

Stacy Rand: Member of the *Wayside Station* Merchant Guild.

Lieutenant Jackson: Commonwealth Navy Security.

Rear Admiral T. S. Melnyk: Commonwealth Navy Security.

Marion de Luca: Remus' Italian descendant.

Marcelo: Englishman immigrated from Italy to England.

Cambridge twin brothers: Miners.

Jonathon: Mark's son.

Cheri: Niece of the Cambridge brothers and wife to Jonathon.

Doctor Shan McGee: Civilian doctor.

Commander Steiner: Search and Rescue.

Captain Tom Steiner: Search and Rescue.

Sergeant Miller: Marine security for Search and Rescue.

Glossary

Drive Pile: Mixture of rare metals that hold the power to operate spaceships.

Flitter: (Old Navy terminology) **F**rigate, **L**ightweight landing craft, **I**nterplanetary, **T**ransfer personal, **T**ransport, **E**mergency anchorage and **R**econnaissance vehicle. The flitter uses liquid fuel for the fuel cells instead of a drive pile.

G-Gun: **G**enocide systematic cellular destruction Gun.

Planet Zaroma

Natives:

Asoma: Standby Elder.

Carville: A female native, also known as the *Wise One.*

Pica: Weaver, also Chantal's friend.

Norick: Pica's mate.

Marica: Healer.

Kon: Family friend.

Pepa: Kon's coon.

Villages:

Vica, Paca, Nica, Mica and Whiz.

Geographic Features: Vica Canyon, Nica Valley, Mica Valley and Canyon, Water Stones, Whiz Valley, the Great Divide and the Great Depression. The Inlet Sea, Hot Spring Lake, "Tranquility," and the Northern Great Lakes.

Planet Earth Timeline

2032-2064. The Great Loss Depression

2115. The First UN-Sponsored Expedition beyond the Solar System

2166. Earth's First Great Exodus

2217. Earth's Second Great Exodus and the Formation of the Imperial Confederation

2271. The Imperial Confederation Forms the Four Sectors

2290. The Terrible Robot Rebellion: A Ten-Year War

2307. Chantal Is Born

2310. Mark Is Born

2322. The Imperial-Thracian War

2332. The Formation of the Commonwealth

2335. The *Marvel's* Auction

2336. The *Marvel* Is Renovated and Launched

2366. Mark Meets Katherine

2367. Mark and Katherine Marry

2368. Crystal Is Born

2386. Katherine Dies

2386. Mark Adopts Misty

2390. Mark Hires Chantal

2392. The *Marvel* Is Grounded on Planet Zaroma

Planet Zaroma Timeline

2393. Second Thracian War

2393. In Year One, Cheri Is Found

2394. In Year Two, Jonathon Is Born

2400. In Year Eight, Misty Dies

2407. In Year Fifteen, the Styler Family Is Kidnapped

2418. In Year Twenty-six, Mark Dies (at age 107 years)

2429. In Year Thirty-seven, Chantal Dies (at age 120 years)

The Solar System

The Commonwealth governs four territory sectors.

Sector One:

- Solar Titus, Planet Chowan
- Solar Academy, Planet Academy
- Solar Alom, Planet Zaroma

Sector Two:

- Solar Sol (Sun), Planet Earth

Sector Three:

- Solar Sunna, Planet New Sweoland

Sector Four:

- Solar Brandon, Planet Rosser
- Solar Stelletta, Planet Remus
- Solar Stella, Planet Romulus
- Void (Wayside Station)

Dedication

To my lovely wife, life partner, and friend. Thank you. Without your support and patience, I would have never achieved my dream.

Acknowledgments

I want to thank my two editors: Maria Mikic, my creative writing tutor, and Robert Cahill. Without their help, this book would never have been completed. Thank you for your patience, guidance, and use of the editor's red pen.

About the Author

Robert David Jensen, a retired Millwright from British Columbia, Canada, discovered his passion for storytelling later in life. His journey as a writer began with a heartfelt biography about his wife Eileen's grandmother, who bravely immigrated to Canada from Europe in 1909.

Inspired by this family tale, Robert turned his creative energy toward crafting his own imaginative worlds. In 2017, he released *The Masquerades*, the first in an enthralling sci-fi series, which was soon followed by *The Prodigal Daughter* in 2018.

This edition marks The Masquerades's second publication, inviting new readers and returning fans to experience the adventure once more. Robert's stories reflect his lifelong dream, bringing the boundless possibilities of space to life while staying rooted in the timeless human experience of discovery and connection.

You may contact Robert at: bookofexcelllence@aol.com

Preface

To my readers,

I have been told that because of my Christian background, I should not write science fiction novels. I'm afraid I have to disagree, but I do recognize that a few of the Christian-themed books I have read are embarrassing pieces of literature. I hope to change this perception.

"The Masquerades," which is Book One of a series of two books, consists of two parts. Part One contains Chapters 1-11 and Part Two contains Chapters 12-20. Each book flows in continuity and has the capacity to stand independently as well as they have different but related characters.

I realize that many people judge a book by its cover and title and good books are often ignored because the reader doesn't like the title or cover picture. The title and sub-title I have chosen are long and simple, and hopefully, they will pique readers' curiosity enough to open the book.

Introduction

We all have secrets hidden somewhere in our past. Many are of little value, while others may be sources of embarrassment. What if your secrets could hurt or harm those closest to you?

Captain Mark Styler is blamed by his daughter and extended family for his wife Katherine's death. Chantal, the Android and her secret fugitive past, and Sara, the ship's computer, will cause trouble for all those who know her.

These three individuals live and work on the *Marvel*, a private merchant ship, each holding secrets that could result in dangerous circumstances. Despite the incredible odds that impact the hopes and expectations of each, "All ends well."

Or does it?

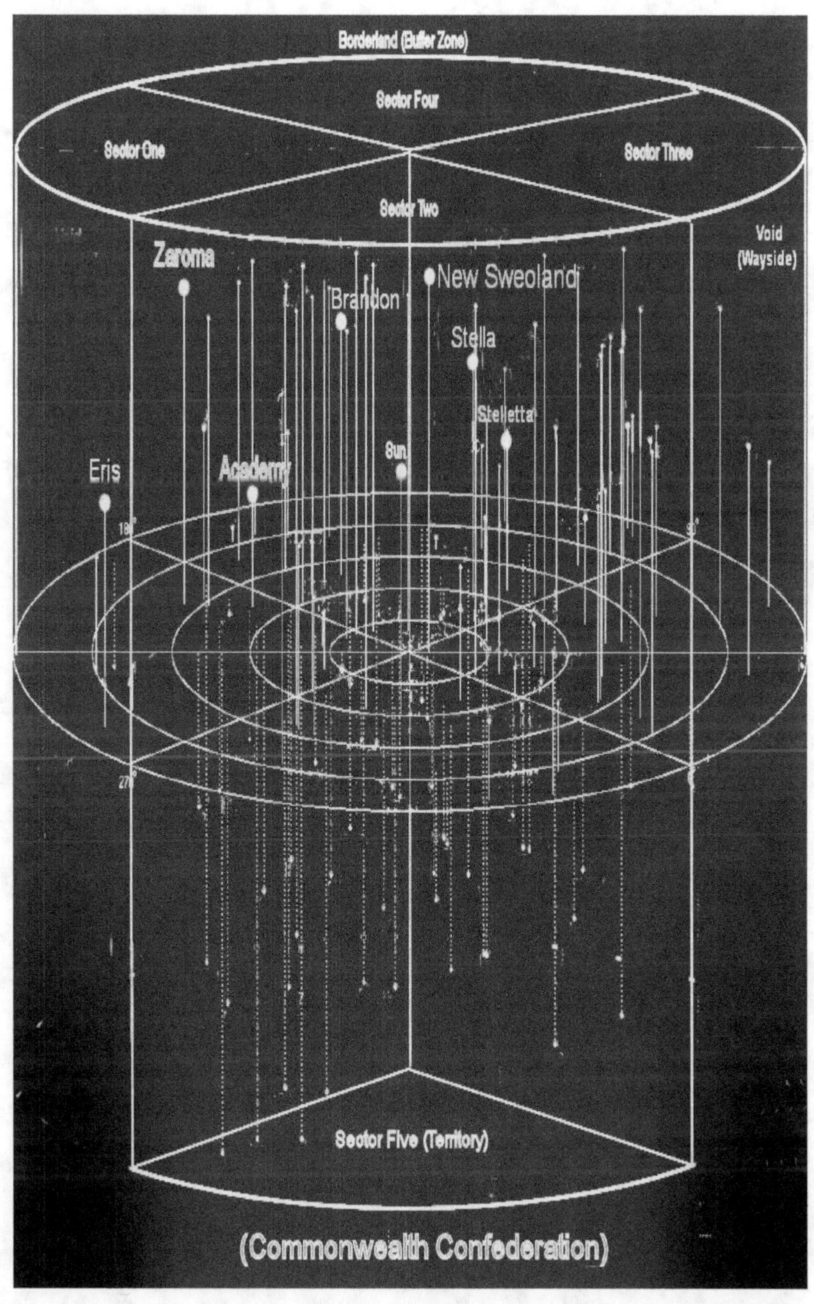

Map One: The Commonwealth Confederation's Four Sectors

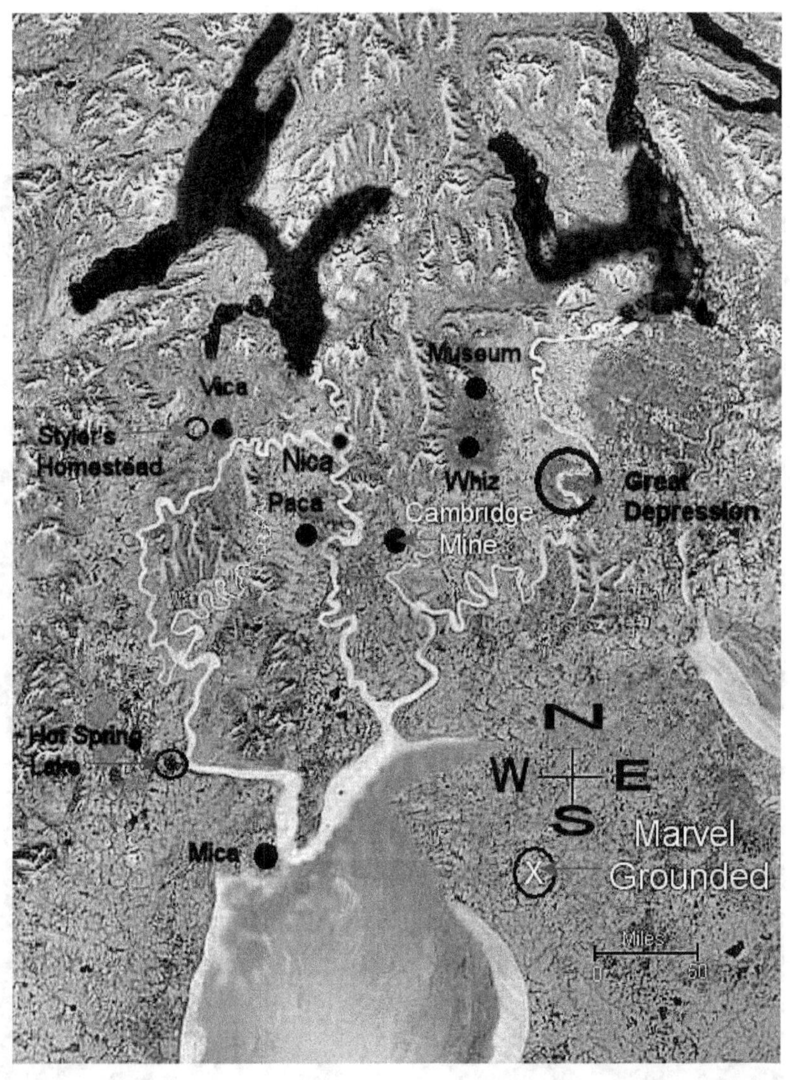

Map Two: Planet Zaroma, Styler's Homestead

Part I

Chapter 1

I am always amazed at how our life's journey may begin or end. A slight change in a given chain of events could have a positive result or potentially lead to disaster. My grandfather always told me, "You are the spitting image of your mother, but you have your father's personality."

My grandfather and father had a love-hate relationship that I could never understand. As I grew older, my father told me, "Find three people that will be a major inspiration for you."

I am happy that my father was the first person and my husband the second.

Then, one day, Chantal became the third.

C.A.S.

The ground taxi crossed the bridge over the river and turned off to the freeway's right. The hidden Sun rose above the edge of a thick cloud covering the spaceport near the city. The taxi made its way toward a red building outside the fenced boundary area of the local spaceport. The building was plain, with few windows.

The taxi parked in front of the main entrance. A woman of medium height with short brown hair, brown eyes and the grayish skin of an android stepped out of the car. She straightened herself and looked up at a sign above the door which read, 'Merchant Trader Employment Center.' The woman glanced around, seemingly puzzled that so few people were in the area. She walked through the sliding door into a small, empty lobby devoid of its natives.

A hologram of an android appeared behind the counter booth.

"Your ID, please."

"Chantal, AN-2134-2," Chantal replied.

"Thank you. Please enter the hallway to my right, turn left and enter the last door to the right."

Chantal proceeded down the hallway, approaching the final door, which slid open noiselessly. The room was large, with twenty vacant administrative desks, except for one. A man with gray skin similar to Chantal's was sitting behind the desk. Androids, synthetic organic entities with limited emotional aspects to their personalities, were perfect for professional positions. All androids in the Commonwealth worked under strict regulations, and most androids provided more than just manual labour. The Commonwealth authority was not to be questioned by anyone. Chantal knew that people disappeared when they asked too many questions.

In a monotonous voice, the android asked, "May I probe your ID?"

Chantal tilted her head to the right to reveal the neon metallic receiver imprint on her neck.

The android used his handheld scanner to touch the imprinted symbol. Satisfied, he set the hand probe back on the desk.

"Thank you," the android said as he looked at the monitor, "My name is Andy," he continued, without looking at Chantal, "You have an interesting employment history. You have both a medical degree and a class one pilot license. It says here that you spent your first twenty years as a standby pilot for a scout mining company. During the last 18 years, you worked as a first-class pilot and doctor for the *Highmore Space Liner*. This employment is most unusual and almost unheard of for an android!"

Andy raised his head to face Chantal, "You are extremely overqualified. All we need is a class two pilot and a first aid attendant. Why would you want to work for us?"

"Different reasons," Chantal replied. "I never liked working with a large crew and passengers. It was interesting for a while, though. I just want to go back to the basics—it seems that I miss that work most of all. I have read about Captain Styler and the profile of this ship, the *Marvel*. Your captain has one of the best reputations among the private Merchant Traders."

"Yes, Captain Styler does have a good reputation, and after ten years working for him, I will miss him," replied Andy.

"Are you going off-ship?" Chantal inquired.

"Like yourself, I think it is time for a change, but I promised the captain I would find a replacement before leaving. Even though you are overqualified, you will be an excellent replacement candidate."

Andy folded his hands and continued, "The wages and

benefits are the standard rates and come with a one-year contract, with an option for a one to three-year renewal term. Do we have an agreement?" asked Andy.

"I do have a question," Chantal said to Andy. "Where is everybody? Shouldn't this place be full of clients and staff?"

"Today is a local carnival holiday, and Captain Styler has a business obligation to complete, which has called him away. That is why you don't see the office at full staff. Even though it is rather unorthodox, we decided to schedule your interview. Are you prepared to accept the contract? I am about to begin the electronic transfer. Are you ready to receive the data?"

Chantal touched her neck implant receiver as the contract was downloaded and stored in her implant. Chantal retrieved the contract, engaged her comprehension chip, and then read it in her mind's eye. Chantal's mind's eye, a primary storage unit, was an implant constructed of crystal that, when installed by a doctor, allowed a being to receive and store electronic data by connecting directly to the brain. With an implant, the user could see a document or images in the frontal part of the brain and make changes before sending the data back to the implant. Otherwise known as a mind's eye, the implants were joints for androids, humans, and other advanced sapiens.

"As you can see, the rules and regulations are similar to your previous space liner employment," Andy explained.

Chantal nodded and asked, "Why is rule #9 in place?"

Andy replied, "Captain Styler is religious, so there will be no swearing on board."

"Religious?" Chantal asked.

"Bible study that he does in his quarters. He usually keeps to himself when he does this."

"And what about rule #27?" Chantal read, "All those employed, including androids, on the *Marvel*, are to conduct their activities professionally and respectably with each other."

Andy explained, "What you do off the ship is your own business, but this is his ship, and we must treat each other respectfully."

"Okay, but why does this rule apply to androids? That is unheard of," Chantal exclaimed. "It's infrequent for an android to be treated with respect by society."

"It's not what you may think," Andy said. "It's not just for religious reasons. You will have to ask the captain about it. Just understand that he has a good reason for it."

Chantal paused for a few seconds and replied, "Alright, when do I start?"

"Tomorrow at sunrise local time 07:00 hours. Report to Bill at the Port 21 lobby."

"What is his last name?" Chantal asked.

"Bill doesn't have the last name. He is Captain Styler's oldest and most experienced crew member. If you need anything, Bill is the one to see."

"Why is the ship called the "*Marvel*"? Chantal asked.

"I'll let you find that out for yourself," Andy replied.

"Okay, I will sign on." In her mind's eye, Chantal opened her security ID, agreed to the contract and redirected it back to Andy.

Andy picked up his folder, smiled and said, "Welcome aboard, Chantal and good luck."

It's odd for an android to smile, Chantal thought as she shrugged and walked out of the building and back to the waiting taxi.

As the taxi pulled away from the building, Chantal looked at the spaceport nearby, where she could see various small and large ships. Most of the aging vessels were privately owned, having been previously sold or downsized by large companies or as government-owned ships. A Commonwealth Navy vessel was parked among them.

Chantal wondered if she was doing the right thing. Suddenly, she experienced an odd feeling of contentment. *Why do I feel like that?* Chantal wondered.

"Is your destination back to your hotel?" the taxi driver asked.

Chantal replied, "Yes, please."

***

It was 07:00 hours when Chantal, dressed in her dark blue space uniform, sat in the hotel lobby near the entrance to Port 21. A small man dressed in plain brown coveralls approached Chantal and, in a raspy voice, asked, "Chantal, AN-2134-2?"

"Yes," Chantal replied.

"Very good. My name is Bill. Are these all of your possessions?" Bill glanced at Chantal's cases sitting at her feet. Without waiting for her reply, he said, "Please follow me."

As Chantal rose to her feet, the two suitcases sensed her leaving, hovered above the floor and followed her as she walked beside Bill down the hallway.

"What do you do on the ship?" Chantal asked.

"A little bit of everything," Bill replied, "Janitor, labourer, ship cargo controller, and chef."

"Chef?" Chantal inquired, "You are a qualified chef?"

"Yes," Bill proudly replied, "We do things differently on the captain's ship."

"Then you have a mess deck with a real kitchen? Not just a self-service eatery?" Chantal added.

"Yep, three square meals a day, and if you touch my kitchen, I will chop off your hand!" Bill smiled.

I must remember to keep that in mind. I'm not sure if Bill is kidding or not. Probably not, Chantal hoped.

"You will be impressed with our ship," Bill smiled, then announced, "Here we are! It's an old ship; a little slow, but completely rebuilt when the captain bought it at the auction fifty-three years ago," Bill said proudly, and added, "All of the crew members have their quarters, which are twice the standard size. It is designed for a long haul without refuelling at every port entrance. The advantage of this ship is that it can land with only a foot of space underneath her, so we don't need a cargo lift. Some of the ports we travel to are somewhat backward." Bill smiled and continued, "The *Marvel* is a rarity. There are not many of them around anymore. She is wide and squat at approximately 100 feet wide and 300 feet long. The *Marvel* consists of two decks. The forward half of the first floor is the cargo haul, and the bridge is above the forward cargo haul. The back of the bridge leads to the center gallery, which directly connects with the captain's quarters, the library and entertainment room, the fitness room, the sick bay, six living quarters, and the mess deck and kitchen. From the kitchen, the gallery leads to the back two-story deck, the repair shop, the fuel cell drive piles that power the anti-gravity lift drive, the sub-light pulse drive and the faster-than-light warp drive to enter warp space."

As they stepped upon the ship's cargo ramp, Chantal noticed that the cargo hull was full of various-sized containers secured to the floor and wall.

"It looks like this ship is ready to go," she observed.

"Yep, the captain is waiting for flight plan clearance from the control tower, and then we are ready to go. Let me take you to your quarters, then meet me on the mess deck and we will wait for the captain to show up on the bridge."

Chantal followed Bill up the stairs from the back of the cargo hold to the second floor.

"How many crew members are employed on the ship?" Chantal asked.

"Including you, one android, four crew members, and the captain. Mark is the ship owner/operator. The captain is the first-class pilot, navigator, and engineer. Mason McDonnell is the chief drive engineer. Tim Andorra is the electrical apprentice. There is me, and you as the reserve pilot and first aid attendant." Bill smiled. "And of course, there is Misty!"

"Misty?" Chantal asked. "What does Misty do?"

"Misty can make you smile. Oh, I forgot you androids can't smile!" Bill grinned and exclaimed, "Turn around and look behind you!"

Chantal turned around and saw a small dog sitting a short distance away. "You have a dog on this ship?" she asked.

Bill replied, "When you get to know her, you will realize Misty is a brilliant dog. Misty is everyone's friend, but she belongs to the captain, so you must be careful what you say in front of her, or you might hurt her feelings."

"How can a dog be that sensitive, and why does Captain Styler call her Misty?"

Bill's smile tightened. "That is a long story, and you must ask the captain that question," he replied.

"Here are your quarters," Bill explained. Breakfast is served at 07:30 ship time. Take a few minutes to familiarize yourself with the ship and meet the crewmembers. The captain has told me that he should be on board at about 09:00."

"Is that what you call him, 'captain'?" Chantal

inquired.

"Sometimes the crew calls the captain 'Mark'," Bill replied, "And sometimes we just call him 'captain'. I suggest you call him Captain Styler unless he says otherwise. Because of the years we have spent together on this ship, we are all like a family."

Chantal was impressed with her new quarters. A real bed, not a bunk bed, a desk, a hidden closet, and a private bathroom, too! She quickly unpacked her belongings and placed her suitcases in the storage room next to the washroom. At 07:30 hours, Chantal remembered that it was breakfast time. She walked back to the mess deck and saw the ship's crew members, who were already sitting at the table.

"Hey guys," said Bill, standing behind the kitchen counter. "I'd like you to meet our official new first aid officer and reserve pilot, Chantal. And guess what? She is a real doctor, too!" Bill bragged.

The crew looked up in surprise as Chantal walked toward the table and stood before them.

A wide-shouldered, stocky man with a black crew-cut rose up and said in a friendly tone, "Hello, and welcome aboard! I am Mason, the chief engineer. This is Tim, the electrical apprentice, and of course, Bill, you have already met." Misty paced over to Chantal, stood on her hind feet and offered her right paw.

"I suggest you shake her paw, or you will offend her," Mason said.

"Are you serious?" Chantal replied. Mason smiled and

nodded his head.

Chantal thought, *I can't believe I am doing this*, as she bent over and shook the dog's paw. "Hello, Misty!" she said heartily.

When Chantal stood, Bill said, "Don't forget Sara!"

Sara? Chantal looked around but couldn't see another crew member.

"Hello, Chantal! It's a pleasure to have you aboard. I am Sara, the ship's computer," said a voice.

A computer with a name, Chantal thought. *It's a long-established tradition in the Commonwealth that ships' computers are never given a name. The reason has never been questioned for fear of being rebuked by superiors on the job.*

Bill said, "Sara is an exceptional computer as far as computers go. Sara can answer any engineering, electrical, navigation, or medical question. Sara can even provide me with recipes," Bill laughed. "Okay, everyone, time to sit around the table. Breakfast is now served."

Chantal was surprised as Bill brought out large bowls of oatmeal, toasted bread, drinks, and fruit. Synthetic food or food paste was available on most ships except for first-class liners to save on space and costs.

Bill sat down, bowed his head and prayed. Chantal was unsure what to think of the rest of the crew and glanced at Bill respectfully. Saying grace was a rarity nowadays. Eventually, Bill raised his head and said, "Okay, let's eat."

Chantal shrugged her shoulders and started serving

herself breakfast.

"Chantal, tell us about yourself," Bill remarked.

Chantal looked up, startled, and stared at Bill. "Why are you asking an android?"

"Why not!" Bill replied. "This crew is not judgmental or indifferent toward androids like some folks out there. We are all going to be traveling together for days, weeks, and months in space. It's good for us to know about each other up front."

Chantal took a single large breath. "Well, to start with, you may call me Chan. The Android Department sent me to the industrial mining outpost near the *Borderline* between the Commonwealth and the Thracian Empire sectors. Afterward, we were herded with other androids and humans on a Navy Search and Rescue ship before the Thracian—invasion destroyed our base. When the Thracians started their destruction of our base, they killed the human pilots. The ship's doctor downloaded the pilot program and stored it in my memory bank. I was the first known android in the mining outpost to be permitted to become a pilot. Our ship arrived at the nearest Commonwealth-occupied solar system, Eris, to escape the Thracian invasion."

"It was a miracle that the ship made it to the next solar system without the craft being destroyed," Bill said. "Under different circumstances, you wouldn't be sitting here."

"I had the knowledge downloaded into my memory bank but not the pilot experience," Chantal replied. "Our craft crash-landed. Without an owner to claim me, I became a free agent and signed up as a co-pilot for another small private mining scout. Our assignment was to deliver geologists for rare metal core drilling operations to minor planets or asteroids."

"Later on, I signed up as a co-pilot for the Navy Search and Rescue," Chantal continued.

"The Navy allowed you to be a pilot?" Mason asked doubtfully.

"When the Commonwealth formed one hundred years ago, it controlled the three Navy divisions. Over the years, political power factions have divided these further. Nowadays, three separate administrators govern the three sectors: the Navy, Navy Security, and Navy Search and Rescue. The Navy Search and Rescue has been the most liberal. Androids have been allowed more freedom there than anywhere else among the Commonwealth's four sectors. I can't explain why; we have just accepted it." Chantal paused then continued, "After twenty-five years of service, I left the agency. Before I left, my commander downloaded a legal first-aid medical program into my mind's eye as a farewell gift. Later, I signed up as a first-aid attendant for a chief executive officer of another private mining company. This CEO owned a personal yacht. After several productive years as a ground crew driver and first-aid attendant, one of the natives attacked us at one of their nearby major mining refinery plants. Afterward, we learned that one of our crewmembers had mistakenly destroyed one of their sacred sites, and the natives took revenge. The CEO and the doctor were both injured in the attack. The doctor needed to download an illegal advanced medical program into my memory bank so that I could perform surgery on them both."

"Did the Commonwealth authority charge the doctor for the illegal download?" Mason asked.

"No, the doctor was too badly injured and died, but the CEO did survive the surgery. Out of gratitude for saving his life, the CEO sent me to a remote star system outpost where illegal programs are ignored due to a shortage of skilled labor. This often happens when companies operate further away from the Commonwealth's main populations and trading routes. When my contract was up several years later, I signed on with a private luxury space liner, the *Highmore*, as a general medical-surgical nurse. The ship's owner didn't care about the illegal doctor program as long as I kept a low profile for the next 20 years. As a free agent, not a bonded servant, I had greater freedom to mingle with the passengers and crewmembers."

"Why did you leave the liner?" Mason asked.

I decided it was time for a change again and wanted a down-sized company to work with. I thought working for Captain Styler would be an ideal new employment opportunity," Chantal replied.

Mason stared at Chantal and her body language for signs of doubt and quietly looked for emotions in the stoic android. Twenty years with one of the most desired career opportunities, and she left her employer? *Something doesn't sound right*, Mason thought.

"It's 08:30 hours, guys. It's time to clean up," said Bill. "The captain said he would be here at 09:00 hours with the release certificate and flight plans to leave off-world."

"Tim, Chan hasn't seen the rest of the ship yet, and you

need to have her on the bridge before 09:00. I will clean up for you both today," Bill suggested.

"Thanks, Bill," Tim replied as he led Chantal past the crew quarters and stopped at the recreation hall stocked with various types of exercise equipment. Tim showed Chantal a library across from the recreation hall that was filled with real hard copy books, as well as an entertainment memories disc and a large table for meetings.

"The next room is the sick bay—you probably will be impressed with this," Tim said.

The door registered Tim's presence, and they entered the room.

This sick bay is equal to any modern space liner. There are two single medical diagnosis hospital beds and a full-sized analyzer monitor above each bed. All necessary, appropriate, professional medical equipment that may be required is clearly in place, Chantal thought.

"I am impressed, Tim," Chantal said, enthused. "Captain Styler has invested a lot in this sick bay! Why did the captain go beyond the standard first-aid clinic required by the Commonwealth medical regulations?"

Tim replied, "You must ask the captain about that." He moved forward. "The next rooms here are the captain's quarters, and over here is the bridge."

"Hi, Sara," said Tim. "Sara, can you take over for me, I have a lot of things to do."

"I will leave you with Sara," Tim said to Chantal.

"Hello, Chantal, or I should say Chan," Sara responded in a mature, pleasant female voice.

Chantal was startled and thought, *This ship's computer sounds too human.* Chantal couldn't remember ever hearing any ship's computer speak in anything other than a monotonous voice. *And ships' computers are not supposed to be designed to respond to spontaneous conversations. How is it that this computer is allowed to exist?*

"Hello, Sara," Chantal replied awkwardly. She passed six empty passenger seats as she moved toward the front of the bridge. She took her place in the co-pilot chair.

"Are you ready for the ship's controls and specifications to be downloaded to your implant?" Sara asked.

Again, the ship's computer spoke without being asked. How very strange! Chantal thought. She touched the printed implant symbol on the left side of her neck. "Computer, you may download the ship controls now," Chan commanded.

Chantal immediately became familiar with all the ship's controls and capabilities as she thought, *Now I know why this ship is called **'Marvel'**! The ship is entirely updated and upgraded with the most modern electronic components, and the engine drives are very advanced for its age. If an emergency happened, the computer could take over on its own! No wonder the ship has a small crew. Where did the captain get the substantial financial backing to pay for all these upgrades? Private independent merchant traders are* *not known for their economic success.*

Chantal proceeded to make sure everything was ready for takeoff. She checked to ensure the drive pile was fully charged, the fuel tank was complete, and the flight and navigation controls were ready. The Commonwealth authorities had a tightly controlled flight route in the four sectors to track all ships moving through space. Flight plans must be made available to the Commonwealth authorities as part of the 'permission-to-travel' process, with an approved certificate released to the vessel's ship owner or captain. All she needed now were the flight plans from the captain.

At that moment, Sara announced, "Captain Mark Styler is on board. I have notified the captain that you are on the bridge and are ready to take direct control of the ship. Captain Styler said he would join you on the bridge and asked that you start the lifting protocol."

Chantal engaged the start-up on her console and could hear the engines engage as part of the anti-gravity lift and pulse drives.

Moments later, a tall, slim man with short brown hair and green eyes, dressed in a black uniform with a slight yellow outline around the waist and collar, entered the bridge.

Chantal rose and bowed, showing the appropriate, respectful behavior expected between an android and a superior. Mark smiled and bowed back, surprising Chantal.

"Sara just told me you prefer to be called Chan instead of Chantal," Mark smiled. He continued, "You may have noticed by now that everything here is very informal. So call me Mark or captain."

Chantal, taken aback by the captain's pleasant manner, said, "Thank you, Captain."

"Are you ready to take flight by 09:30 hours?" Mark asked.

"Yes, Captain," Chantal answered.

Mark sat in his pilot chair and said, "Chan, I just downloaded all the flight plan destinations to the ship's computer. You may now download the first haul on the flight list."

Chantal touched her implant as the flight data was downloaded to her memory bank. "Sector One, Solar System Titus, planet Chowan, 1.2 Earth's gravity. The estimated travel time is 4 to 6 days. Cargo is 2.4 tons of freeze-dried spices," she repeated. "Are we delivering spices?" Chan asked.

"Yes," Mark replied, "The Chowan natives prefer our spices, which are superior to any of their own. The natives can't grow them because most of their planet is arid, and the humidity makes the climate too dry. The profit margin is too good to let this one go."

Mark touched the communication switch on his console. "Bill, is everything in the cargo bay locked down?"

Bill replied. "Yes, Mark."

"Mason, how is the drive pile?"

"The drive pile is ready and stable, Mark," Mason announced.

Tim and Bill entered the bridge and sat in their respective chairs.

"Sara, do you have anything to report?" Mark asked.

"Control tower has announced that the 09:30 hours clearance is on schedule!" Sara replied.

"Thanks, Sara!" Mark said.

Mark engaged *Marvel's* interior-wall anti-gravity null plates, which canceled out the planet's increased gravity pull and the ship's movements through space. The floor null plates compensated for the momentum effect of the onboard weight.

"Okay, Chan, I'd like to see how you do from here. Please take over!" Mark commanded.

Chantal, caught off guard, quickly responded, "*Marvel* to control tower, ready for 09:30 hours lift-off."

"Control Tower to *Marvel*, you may proceed to lift off!" came the authorization from the tower.

Chantal engaged the lift drive. Under the ship's belly, the thrust's white glow lifted the *Marvel* 1000 feet into the atmosphere.

Chantal engaged the main pulse drive engine, and the ship slowly moved forward and then picked up speed as she moved into a high planet orbit. Before entering warp space, Chantal locked onto the safe zone point for the warp space entrance, a gravity-neutral area free from any nearby planetary influence. An enormous gravity influence could cause a ship entering warp space to wander off from its final destination course.

The *Marvel* moved up to quarter light speed. Chantal turned to Mark and said, "Warp space jump is about 30 minutes away."

Mason entered the bridge and sat down in his chair behind Mark. Referring to Chantal's pilot skills, he asked, "How did it go, Mark?"

"Not bad, Mason," Mark whispered. "I think Chantal will be a reliable pilot. We will see how she does when we enter warp space. I will be back in 25 minutes."

"When the bridge door had closed behind Mark, Bill spoke up. "Hey, Chan, that was smooth and quick. You are as good a pilot, just as the captain!"

"The captain is not upset, is he?" Chantal inquired.

"No, Captain Mark wanted to see how you handled the stress factors involved in take-off. Mark has done that before with other pilots. You're okay!" said Bill as he left the bridge.

Since there wasn't much to do for the next 30 minutes and Chantal had become aware that she was alone with Tim, she asked him to explain how he had become a member of the crew.

Tim began, "My dad once worked for Captain Mark. Dad left shortly after the captain's wife, Katherine, died."

"How long ago has it been since Katherine died?" Chantal asked.

"About four years." Tim replied and continued, "Three years ago, Dad asked Captain Mark to take me on as an electrical technician apprentice and trader merchant. That's been my dream ever since I was a kid, to be like Dad, traveling in space and seeing the universe."

Chantal suggested, "That must have been hard for you and your mother with your father away for weeks or months at a time."

"My mother died when I was a baby, and my mother's sister and her husband took care of me. Dad stopped by whenever he could. Dad is a storyteller, so he has always had a story to tell me, mostly about where he traveled and worked," Tim replied. "After Katherine died, Dad decided to retire and marry his childhood sweetheart."

"How old are you now, Tim?" Chantal asked.

"I am twenty-three, and I will have completed my apprenticeship in one more year. Then I will go back to school for the advanced tech program!" Tim smiled.

"It must have been a big favor for Captain Mark to take you on board," Chantal said. Tim looked down at the floor and then turned his head aside to the right as he stood up. You will have to ask the captain about that one. I will be back in a few minutes," he said as he left the bridge.

This is the third time I have been told to ask the captain about questions I have! Why the secrecy? Chan thought.

Abruptly, Misty appeared, wandered onto the bridge, jumped into the pilot chair and sat staring at Chantal. Chantal stared back at the dog. *There is something odd about that dog!* she thought.

Tim returned and sat down. "Hi, Misty. Do you want to sit on my lap?" he asked.

"What is that dog doing here?" Chantal asked.

"We don't refer to Misty as a dog; she is just Misty!" Tim replied. "Misty belongs to the captain, so she can go anywhere on the ship. She is astute. Sometimes, I think she can even understand what we are saying."

"Why do you say that?" Chantal asked.

"It is just how Misty looks at you when you say something," Tim answered. "I was hoping someday to have one of Misty's puppies, but the captain told me she was sterilized."

"Where did the captain find Misty?" Chantal asked.

"I don't know, and the captain wouldn't tell me when I asked him," Tim replied.

Sara announced, "Warp jump in 5 minutes."

Moments later, Mark and Bill arrived on the bridge. Bill took the back seat, and Tim slid forward to his console. Mark sat beside Chantal, staring ahead at nothing in particular, then collected his thoughts.

"Chan, prepare us for warp jump," Mark commanded.

"Mason?" Mark asked.

"Warp drive's ready," said Mason.

"Sara, open the warp space fabric."

The main forward viewing monitor showed a circular rainbow halo, the open barrier between normal space and warp space.

In the early years of space travel, it took many attempts at warp jumps to reach a single destination. Over the years, the Commonwealth installed homing signals in every solar system in the four sectors that spanned 50 light years, otherwise called sensor probe markers.

"The sensor probe marker is now registered, and the navigation system is locked on," said Sara.

"Estimated time, Sara?" Mark asked.

"The Estimated time in warp space is 3 days, 2 hours, and 20 minutes. Fuel consumption is estimated at 12% to our destination," Sara intoned.

Chantal asked Mark, "The ship's computer can calculate the exact time?"

Mark smiled at Chantal. "Sara is excellent in her estimation!" he said proudly.

As she returned to the images at her console, Chantal didn't know what to think. Even the computers in the Navy or space liners she had once worked for couldn't do better than give a day's estimate! *This computer couldn't be the original ship's computer,* she thought. *Another question to ask the captain later!*

Chan engaged the warp drive. The *Marvel* became a long streak of light, then vanished into warp space.

Chapter 2

"Five, four, three, two, one... we are now out of warp space. The *Marvel* is now in regular space," Chantal announced.

"Sara, our location, please?" Mark asked.

"The scanner shows we are on course," Sara replied.

Are we? Chantal thought, *Is this pure luck? The ship's computer's calculation capabilities are incredible!*

"Captain, our estimated arrival time at Planet Chowan's orbit entrance is 1 hour, 14 minutes." Sara added.

"Alright, I am going to be in my quarters and will be back in 1 hour." Mark rose from his chair and left the bridge with Misty tagging behind him.

Chantal turned to Bill and asked, "Does the captain always do this at the last minute? Normally, when a ship is near a planetary vicinity, the captain stays on the bridge until landfall."

Bill sighed and replied, "Mark has had difficulty adjusting to his loneliness since Kate died. It is just his escape mechanism when he needs to be alone."

Chantal shrugged and turned back to the ship's monitors. "Let's talk about you, Bill. How did Captain Styler come to employ you?"

"Well, that started on Planet Rosser from the Brandon solar system in the third sector about 53 years ago."

"Did you get that job through an employment agency?"

"No!" said Bill. I got the job through an old girlfriend. I was a nobody, down on my luck. I had no trade to speak of, but I loved to cook. I did all kinds of short-term manual jobs just to make a basic living."

"Then I met this girl, Theresa. She was one of the spare dispatchers at the control tower. Rosser was a pioneer trading post, so there wasn't much space traffic then. Theresa took a liking to me when she learned from one of the ground crew that I was an orphan. Theresa was short and slim like me, with blue eyes and long black waist-long hair. In the beginning, we were just friends," Bill explained. "Before long, there wasn't much trading business at the port, maybe two or three ship visits a week, so my boss cut me loose. Mark and his ship arrived on Rosser one day. He was looking for a new crew member. Theresa asked Mark if he would hire me. Mark agreed, and I have been here ever since."

"I'm curious," Chantal asked. "Whatever happened to Theresa?"

"Oh, we still keep in touch," Bill replied. "A few years later, Theresa found employment at Wayside Station as a dispatcher. Today, she is their chief central cargo coordinator. We both decided soon after Theresa became employed at Wayside that a committed long-distance relationship suited us just fine."

"Where is Wayside?" Chantal asked.

"Wayside Merchant Trader Station is in the 'Void' near the Borderline."

As Chantal recalled, the Void was located in Sector Two, an area in which there were no stars or planetary objects within a ten-light-year distance from the nearest star system.

"I should have asked you when we first met. Why don't you have a last name?" Chantal inquired.

Bill replied, "Shortly after I was born on Rosser, my parents abandoned me at a clinic in a ramshackle town. They didn't know what to do with me because neither the clinic nor the town administrators were set up for baby placement. A local couple from the outskirts of the town took pity on me and agreed to adopt me. They raised me as if I were a valuable slave and Bill was the only name they gave me."

"I ran off with a friend in my teens and found employment at the Rosser spaceport. For the next twenty years, I worked as a fuel assistance ground crew member, as a cargo hauler, and in other manual labor areas. After ten years, I happened to be at the local library one day when I first met Theresa."

An odd beep came from the front console, and a black disc slid out.

"Chan, there is a coded message for Mark from the orbiting satellite," said Sara. "Could you please deliver it to him?"

Chantal picked up the black disc and strode to Mark's quarters. Mark was sitting at his desk reading when the room's computer announced that Chantal wished to enter. "Open," said Mark in response and looked up.

The door slid open. Chantal said, "I have just received a coded message addressed to you," as she handed the disc to Mark.

Mark looked over the disc's contents while Chantal gazed around the room. She noticed many interesting artifacts that Mark must have collected in his travels to various planets.

"You have quite a collection here!" Chantal remarked as she studied the different objects.

"Thanks," Mark replied, "Most of them are gifts from my regular clients. A few I bought," he smiled.

Chantal could see a guitar hanging on one wall of the room.

"Do you play the guitar often?" asked Chantal.

"I used to play before Kate passed away," Mark replied regretfully.

Chantal noticed an open page of a black book on the desk in front of Mark.

"What are you reading?" Chantal asked.

"This my Bible," Mark replied.

"You have a Bible?" Chantal asked. "Why do you read the Bible?"

"It gives me peace of mind when I need it," Mark explained.

Chantal suddenly felt tense in the presence of the holy relic. Mark's Bible brought back long-forgotten and terrible memories as she fought to control her rising panic.

"Are you not well?" Chantal asked carefully.

Mark smiled and explained further. "I am fine," he said. "This Bible originally belonged to Katherine. After she had died, I was going through her things and found a note inside the front cover. She wanted me to have it in case something ever happened to her. I have been reading the scriptures ever since."

Chantal needed to leave the room and replied, "If you don't mind, Captain, I should return to the bridge."

Mark nodded his head as Chantal left the room.

As Chantal returned to the bridge and her co-pilot chair, the vision of Mark's Bible was fresh in her mind. She tried to rationalize various triggers of old memories: a human mob yelling, people screaming, buildings on fire, and death. Above all, she retained a memory of a man with a Bible held high above his head screaming, "Kill them, kill them all!" As Chantal sat musing, she lost all track of time until Tim called out, "Hi, Chan!"

Chantal was startled out of her reverie. She realized that she was daydreaming and quickly composed herself.

Tim was standing beside her station with a notepad in his hand.

"Do you have something for me?" Chantal asked.

"Nothing," Tim replied playfully, "I thought Androids didn't daydream."

Darn it; I mustn't let this slip by, Chantal thought. "You are quite right; androids don't daydream. I mentally checked lists for the next land side protocol coming up very soon."

"What are you doing with that notepad?" she asked.

"Oh, just taking notes recording my daily observation on the ship's electrical read-outs. Sara suggested it as part of my apprenticeship."

At the same time, Mark appeared on the bridge, saying, "There is a slight change of plans. After we finish unloading our cargo, we have other business to attend to below."

"What is that?" Chantal asked.

"We have a new passenger," Mark replied.

"Who is he?"

"She, I think. Or, for that matter, a *Meloson*."

Are you serious? Chantal thought in astonishment and remembered.

Two years ago, an unknown alien spacecraft was sighted in the uninhabited Meloson solar system by company mining scouts, who presumed that the ships were non-Meloson. For some reason, the Commonwealth authorities tried to suppress the rumor of the appearance of the alien craft and the story that spread throughout the four sectors. Chantal remembered that while the crafts were not indigenous to the area, they were called Meloson ships because they had appeared in the Meloson solar system.

Since then, these same advanced alien crafts had also discreetly appeared in other solar systems. The Meloson ships refuse to identify themselves or surrender to the Commonwealth Navy. A repudiation response by a known or unknown craft entering the sphere of a planet or nation is considered a threat.

The Commonwealth Navy couldn't corner or capture the alien ships. Even when fired upon, the aliens and their ships could outrun any missile and shield against the most substantial Navy power beams. The public didn't know much else about them.

"How do we know she is a mysterious Meloson, and why has this ship been chosen?" Chantal asked.

"My contact didn't say much except that the Meloson are very generous with the Commonwealth's credits and seem to pull diplomatic strings by carefully and selectively choosing smaller human-based agencies to carry out their tasks for them discreetly."

"As soon as we unload our delivery, I will interview this Meloson before she steps aboard. If I am satisfied with her answers to my questions, we will accept the contract."

"Where are we accompanying this Meloson?" Chantal asked.

"Our passenger will answer that question when she steps onto the *Marvel*."

"Sounds like she is hiding or running away from something," said Mason.

"I said the same thing to my contact, who assured me it wasn't the case. And the pay is good! Listen closely, though, everyone. Everything must be kept a low profile," Mark said firmly.

"It still sounds a bit risky," said Bill.

"According to Sara, none of the Meloson ships had ever shown any aggression toward anyone, even when their ship was under attack by the Navy."

"Landside will begin in five minutes," Sara announced.

Chantal noticed that the planet below was a desert landscape with a circular greenbelt near the northern and southern poles. The *Marvel* began the descent toward the south greenbelt below. The spaceport appeared at the edge of a massive red desert that adjoined the Greenbelt. The *Marvel* docked near the warehouse trading center.

The red desert reminded Chantal of the Mars surface from many, many years ago.

***

"Alright, that was the last of the containers, Mark."

"Thanks, Bill. We should have our guest and her luggage aboard soon."

"I am hot and sweating. Time for me to clean up!" Bill remarked gleefully, wiping his brow.

Bill was about to pass Mark up the low ramp but abruptly stopped when he saw a metallic-black hooded and robed figure moving toward him. The stranger glided past Bill and approached Mark, followed by a floater, a small anti-gravity platform carrying orange cargo. It passed Bill and moved toward Mark with a strange, soft movement. When the presence paused before Mark, he became suddenly aware that it was at least a head taller than himself. Surprisingly, he wasn't anxious.

Mark looked up, but all he could see under the hooded robe was a visor and mask containing what appeared to be a twin-filtered breathing apparatus. The exposed five-digit gloved hand was made of the same material as the long robe.

A soft voice spoke through the filter mask, "Captain Mark Styler, I presume?"

"That is my name," said Mark. "And yours?"

"You could not pronounce my name, so please call me Melissa."

"Okay, Melissa. Welcome on board the *Marvel*. How may I assist you?"

"I wish to be taken to a place 1.11 light years away from Eris in the same direction as your next haul so my people can arrange a rendezvous with me. We discreetly hired various human investigative agencies and search parties to deliver missing items belonging to my people. There is a possibility of a delivery contract for you in the future. I will explain our purpose after we leave Eris behind us."

"Why this ship?" Mark asked.

"You have been chosen mainly because of your reputation amongst your clients. Your commitment and honesty are well-known in Chowan. You and your crew seem to be very trustworthy. We will, of course, pay you in advance as per your regular agreement, which is what your other clients have suggested."

"What is in your floater, and why is there a low-profile passage arrangement?" Mark asked.

"This trip involves nothing illegal. There is nothing that a port scanner would pick up. As for the passage itself, I have proper travel clearance. Nothing I could say now would satisfy your concerns without jeopardizing my existence with the Commonwealth. I can only explain the mission further after the contract has been approved and you have decided to provide me safe passage to my destination. All I can say is that my intention for this journey is honorable. You have to trust me just as I have to trust you."

"I have one last question to ask you. How do you know about our next delivery schedule?"

"One of our agents overheard your conversation about your next assignment," Melissa replied.

Relying on his instinct, he felt that the Meloson was explaining honestly. Mark agreed, "Alright, I will take you on as a passenger."

"Excellent. Our transaction is now complete."

Mark was surprised at the speed of the exchange and doubted that Melissa could have arranged the transaction so quickly. He quickly asked, "Sara?"

"Credit payment has been transferred." Sara replied understanding Mark's question.

"Okay," Mark sighed, not understanding the technology that would allow such a quick transaction but accepting Meloson's more excellent knowledge. "Please follow me." Mark led the Meloson to the second deck and the passenger reserve quarters.

"Here are your quarters. I have secured your floater in the cargo hold. Standard protocol requires that all personnel and passengers attend the bridge before liftoff. We are scheduled to leave in 10 minutes."

"Thank you, Captain Mark Styler. I will be on the bridge in 5 minutes," the Meloson replied.

Mark turned and strode down the aisle, murmuring, "Now that was one strange lady, or whatever she is." Back on the main deck, Mark moved past Bill and Tim, who were waiting to hear the latest news about the passenger. Mark sank into his pilot chair.

"So, who is this guy?" Bill asked.

"From what I gather, she is a female Meloson," said Mark.

"A Meloson! Wow, a real Meloson?" asked Tim. "What is she like, and what is her name?"

A delightfully mellow voice spoke from the back of the bridge, "My name is Melissa."

They all turned around. A tall, slim alien stood before them in an elegant, long, blue metallic dress that fell to her ankles. Melissa's appearance was very different from what Mark expected. She had a narrow, humanoid face, silky smooth baby blue skin, yellow eyes, and a thin nose and mouth. She was beautiful. The entire crew was speechless.

Her extended hand and arm moved slowly and gracefully as she beckoned the men with five finely boned fingers.

"I thank you for taking me on board. Where may I sit?" Melissa's bell-like tones held everyone transfixed.

Tim stood up and cleaned the seat behind him with his hand. "You can sit here if you want," said Tim.

Melissa softly clasped her hands together and, with a slight bow, said, "Thank you, Tim."

After Melissa seated, Tim asked, "How do you know my name?"

Melissa smiled and said, "I researched the crew members before I boarded the ship. It is our custom to know and understand the personalities of the ship crewmembers so that I won't unintentionally offend any of you or your culture."

"It's time for us to leave. Please secure yourself," Mark announced. Mason?" Mark commanded.

"On my way to the engine room!" said Mason as he hurried from the bridge.

Moments later, "Ready to go, Mark!" Mason announced.

"Where are we going, Melissa?" Mark asked.

"Here is the data, Sara," Melissa said as she flicked her index finger from the thumb toward the control console.

"Data received and downloaded," said Sara.

Chantal and Mark looked at each other with blank expressions.

"Mark, I didn't register any electronic transmission," said Chantal.

"Take us up, Chan," Mark smiled as he gestured with his hand.

The *Marvel* slowly rose above the atmosphere before quickly disappearing into the stratosphere. Arriving at the safe warp jump area, the rainbow opening appeared before the ship, and the *Marvel* quickly entered warp space.

"Melissa, much time will elapse before we reach your rendezvous destination. Do you wish to rest now or would you like to meet in your quarters to negotiate the next assignment?"

"Now would be just fine, Captain Mark Styler."

Mark stepped aside and said, "After you, Melissa."

Mark couldn't help but appreciate her graceful movements and manners. Returning to her quarters, Melissa sat in the nearest chair. She placed her fingers together and rested them on her knees. Mark turned the other chair around and put his arms on the backrest.

Mark observed Melissa thoughtfully. He was fascinated by different intelligent life forms. By human standards and perspectives, he usually needed some time to adjust to the features of most aliens. However, Mark had no problems adjusting to the Meloson and smiling to himself; he thought Melissa's appearance was quite pleasant.

Mark opened their initial conversation diplomatically and politely. "Sara told me that you are half of my weight, so I guess you are from a low-density home world. I assume that part of the reason for the mask and filter is because the air and humidity on Chowan are too hot and dry for your skin."

"You are correct on all accounts, Mark," Melissa smiled. "My world is a great distance outside of your four sectors, so it wouldn't do you any good to know the whereabouts of our home planet. Our people are nomadic in a sense, and that is why I am here. We do need your help, Mark."

"Your technology seems so far beyond us, and yet you need my help?" Mark exclaimed in surprise.

"Your people have done very well in your short three hundred-years in space or, to use your terminology, 'Off-World.' Our 'off-world' amounts to thousands of years, so our technology is far beyond that of your Commonwealth, Captain Mark Styler. However, we have our rivals just as humans do. Your people won the major conflicts against the Thracian Empire beyond the borderland sectors. However, we lost the final battle with our rivals."

"Does that mean the Commonwealth is in danger from your enemies?"

Our rivals are located far from here and are thinly scattered throughout our sector system", Melissa explained." It's doubtful that the Commonwealth will ever cross paths with them."

"So why do you need my help?"

"We are passing through your four sectors en route to our final destination, a star system and the planet suitable for our people. The new solar system has a planet almost identical to our home planet with the same radiation spectrum output of our Sun."

"What is stopping you from leaving?" Mark asked.

"Before the final destruction of our home planet, ten masterships were constructed to transport the last of our people safely away. The mastership I was on was attacked and ransacked. They stole some of our drive piles. The drive pile is required to power our warp drive. Our mastership is crippled, which has stopped us from reaching our final destination. We fought the raiders back to their ship and disabled it. After we had disabled the Raiders' ship, our enemies placed the drive piles within four escape pods. We have recovered some of the drive piles for the last two years by tracking the escape pods throughout the solar system."

"What does that have to do with me?"

"Your Commonwealth authority knows that the Meloson are looking for something, and many of our human agencies are now known. Therefore, we must start looking afresh with new agencies and transporters. Let me show you what I recovered from the last planet." Melissa went over to a small box on a side desk dresser and carefully, with two hands, removed a black stone object the size of a grapefruit. "This is one of our drive piles."

The black stone fit neatly in Mark's palm.

"What is it?" Mark asked.

"An energy transmitter that works only in warp space. Depending on the size of the ship, one or more black stones will allow a ship, after entering warp space, to travel at warp two, which is ten times faster than warp one. Just one of these black stones will power the drive of your Navy's carrier. We need ten stones to power our mastership. They seized four of our black stones. We have recovered two stones, and we still require two more."

"Why don't you just make more of these black stones?" asked Mark.

"The rivals on our home planet destroyed the manufacturing processors. We don't have the time or the resources to make another pile." Melissa handed Mark a miniature black stone embedded into a ring from the box. "This will tell you if this pile is real or not. Place this ring on the larger stone, and you will feel a pulsing vibration."

Mark felt the ring pulse and nodded. "I agree that this contract could be beneficial, but why should your business be conducted secretly?"

"As you already know, many political factions within your Commonwealth authorities and their Navy departments are increasingly under pressure for power domination. The cultural hegemony from many of Earth's ethnic groups that have migrated among many different solar systems is becoming a significant source of ideology in the power structures within the Commonwealth. The Commonwealth is also obsessed with the need to manage advanced technology and the populace to ensure stability, which sometimes causes them to make irrational decisions. Our people are an unknown factor for the Commonwealth political structure, who would, therefore, take a dim view of our contract agreements with private carriers. The risk for you and your ship are minimized if the business is conducted under a low profile. So far, none of our former agencies have been prosecuted."

If we could reach an agreement for each of the two remaining deliveries, 5000 Commonwealth credits would be paid up front, and 5000 upon completion would be deposited into your account at the completion of each contract."

Another 20,000 credits, and I could almost retire comfortably, Mark thought. "Okay," Mark said. "We have an agreement, Melissa."

"I am glad we can do business together. You may keep the ring. We will contact you through it when you are wearing it. You will find that it has multiple purposes, and communication is one of them."

Melissa took the black stone and placed it back into the box.

"Now, if you would excuse me, Mark, I wish to communicate with my people."

"In warp space?" Mark cried out in surprise. "The Commonwealth would give anything for that knowledge," he added.

"The Commonwealth will have to learn that for themselves," Melissa smiled back at Mark. "They are very close to obtaining that knowledge."

"I know this is an understatement, but you know much more about humans than you are letting on, don't you?" Mark momentarily added, "My father was an important businessman, and he told me that client trust must be earned when conducting business." Mark stood up to leave, stopped at the door, and turned around. "You know, Melissa, if the rest of your people are like you, our acquaintance will be quite enjoyable."

Melissa silently bowed her head.

***

"Warp jump out 5, 4, 3, 2, 1."

"Sara, report!" Mark asked.

"We are completely alone," Sara replied.

"My people will be here in a moment," Melissa said from the back row.

"Mark, the scanner shows a huge warp opening just 5 miles away," Chantal reported.

A circular rainbow appeared, one larger than anything Mark had seen before. Then, an enormous ship shaped like a golden sphere dropped out of warp space.

"Chan, how big is that thing?" Mark asked.

"The ship's length is about 1 mile," Chantal answered.

"Why is the ship so big, Melissa?" Mark asked.

"Our ship is an ark. Because we were slowly losing the war, ten arks were built to accommodate our people in suspended animation because of the distance involved."

"How many are asleep?" Tim asked.

"2.5 million," Melissa answered.

"That gives me the creeps; it is like one big tomb," said Tim.

As the *Marvel* drew nearer to the sphere, many hexagon-shaped patterns covered its exterior, and a much larger hexagon circled the middle of the sphere ship.

"I don't see any sign of exhaust. Where is the drive exhaust?" Mason asked.

"The smaller hexagon pattern you see is the drive exhaust. One of the larger middle circular hexagons will open to allow your ship to enter inside," replied Melissa.

Moments later, one of the middle hexagons began retracting in a circular motion.

"Take it slow, Chan," warned Mark.

The *Marvel* entered a vast chamber that contained docking facilities for ships of various sizes and shapes.

"Mark, there is an atmosphere inside this chamber," said Sara. "It seems that an unknown force shield is, therefore, maintaining the atmosphere within the sphere."

"Please dock the *Marvel* where you see the blue flashing light," Melissa instructed.

"Go ahead, Chan," Mark commanded.

"Mark, a traction force has secured our ship," said Chantal.

A walking platform extended from the wall and attached to the side exit door of the *Marvel*.

"This is where I leave you," said Melissa.

"I will walk you to the exit," Mark said as Melissa moved to accompany him to the door.

Her personal cargo floater waited at the exit. *Marvel's* exit door opened, and then Mark and Melissa turned to each other.

"Will we meet again?" Mark asked.

"Perhaps, shortly." Melissa flicked her finger at Mark, "I just downloaded to your implant the instructions for your ring. The ring is registered to your DNA and no one else's."

Melissa and the floater exited the *Marvel* down the walkway of the platform. The alleyway retreated behind her as she walked toward a hexagonal door at the end of the wall and the door closed behind her. The *Marvel's* exit door also closed and Mark returned to the bridge.

"Take us out of here, Chan," Mark instructed.

The tractor force released as the *Marvel* pulled away. Soon the ship and crew were outside the Meloson sphere. The Meloson ship entered warp space, and the rainbow opening disappeared.

Bill asked no one in peculiar, "Now, do you ever think we shall see that again?"

Only time will tell, thought Mark. *Something tells me, though, that the future is going to be rather unexpected and interesting*, he thought as he grinned to himself.

"Sara, calculate our next stop," Mark requested.

***

Chapter 3

Pulling a night shift, Chantal was alone on the bridge when Mason walked in with two steaming mugs.

"I have a hot dark roast coffee for you and a hot Earl Grey tea for me," said Mason.

"Thanks. You're up late!" Chantal accepted the cup and took a tentative sip.

"I couldn't sleep. I'm still thinking about that Meloson ship. When I signed up with the Navy, I spent 36 years learning about all the drive systems in the four sectors, both the Commonwealth's and others'. I would love to study the Meloson ship drive, too. It's been six months, and it is still fresh in my mind," Mason explained.

A companionable silence filled the bridge.

"Why did you leave the Navy?" Chantal asked.

"Stupidity on my part! At the time, I was a chief drive engineer." Mason stretched out and crossed his legs. He took a sip of tea and continued his story. "I had a relationship with a girl named Casey. Casey was a green-horn half my age, a service tech. At that time, the Naval fraternization policy prohibited personal relationships between a chief engineer and enlisted members. We took it quietly, though, and had two excellent years." Mason ruefully smiled, "We had planned a future together." He took another sip of tea, lost in thought.

"So what happened?" Chantal asked.

"An arrogant elite officer, Yung, was enlisted and came onboard the ship. It quickly became known that he was a bully and a womanizer. It wasn't long before the ladies onboard kept their distance from him. He had his eyes on Casey, too."

"One day, Yung was waiting alone for Casey in the ship's passageways. He had pushed Casey up against a wall, and she was scared out of her wits. That's when I arrived on the scene. I took one look at what was happening, went into an uncontrolled rage and slugged him. Unfortunately, he was a superior officer. And slugging an officer means a court-martial," added Mason wryly.

"But you had Casey as a witness!" Chantal exclaimed.

"No, I didn't," said Mason bitterly. "You have to remember, we were dealing with an elite officer and his reputation as well as the reputation of the Navy were both at stake, and they always protect their officers. The investigative committee told Casey that reporting him might compromise her future career. Some things never change, as there was a lot of corruption within investigative committees at that time. For Casey's sake, I told her not to jeopardize her career and to say and do whatever they wanted at the hearing. I was on record as an irrational chief drive engineer. I was given a choice by the investigative committee to accept a voluntary discharge or face a court martial. Because of the fraternization policy weighing as evidence in the court-martial threat, the charges were stacked against me. I chose to take a voluntary discharge to keep my pension and my name off of the court martial list. For the next six months after I left the Navy, I was one angry man! I picked fights, traveled from one town to another, wandered the streets, fueled by the shock of losing my career, my girl, and the unfairness of it all." Mason took a deep breath looking into space.

After a moment, he continued, "One night, I picked a fight with another bitter man—Mark. I collided with Mark as we tried to enter a restaurant simultaneously, and we had a heated argument that led to a fight. It turned out we were evenly matched physically, and ultimately, we both fell to the ground, totally exhausted. As we lay there panting, Mark asked me what had got me so 'ticked off'. I told Mark about my screw-up, and then he said, 'Let's go inside and order a meal.' Sorry, Chantal, I promised Mark his story would be kept untold. To make a long story short, we realized the irony of our predicament. We both laughed as we realized how foolishly we were behaving. Before we left, to my surprise, Mark offered me a job. He needed a replacement for a ship engineer. I accepted the job offer and have been on board for almost four years."

Mason rose. "Time to catch up on my sleep. See you in the morning, Chan."

"Goodnight, Mason, and thanks for sharing your story with me!"

***

"Hey, Mason, give me a hand, will you? This one doesn't have a floater." Mason grabbed the handle on the other side of the container and helped Bill carry it up the ramp into the cargo bay.

Tim entered the cargo hold and said, "Hey guys, after you two finish, the captain wants all of us on the bridge."

"Our jobs are almost finished," said Bill.

When all the crew were gathered on the bridge, Mark announced, "We won't be leaving here for another two or three days. You are all granted a three-day leave." Mark continued, "Another client has contacted me, and I will meet up with her."

"She?" Bill teased.

Mark sighed, "She is an important client."

"Don't you think one of us should be coming along?" said Bill. "This outpost is a rough neighborhood. You usually have someone to accompany you."

"My client made it clear that I am to come alone. If there should be any problems, I will contact and inform Sara. Enjoy your three days off."

Chantal spoke up, "Captain, Bill is right. You shouldn't be alone."

Mark was surprised by Chantal's concern. "Tell you what, Chan, if there is a problem, I will have Sara inform you first."

"You're the captain," replied Chantal, respectfully.

"I am off and will see you all later. Enjoy your layover!"

Mark rose from his chair and quickly exited the *Marvel*. Walking beyond the spaceport, he took a ground taxi and headed toward Francia, a French descent colony. Many of Earth's ethnic society groups during the Second Great Exodus migrated to different planets to preserve their language and culture.

The taxi dropped Mark off in the old section of the town, and he walked along the narrow, cobbled streets. Many of the two and three-story buildings were highly decorative, with ornate architecture, black ironwork balconies, stair railings, and gates seemingly patterned after the classical period of 17ᵗʰ Century France. Mark knew the Old French district was popular with local and off-world tourists. He observed a lively community with many inhabitants and tourists congregating around small tables in front of the sidewalk cafes as he walked along.

The morning's high humidity in the area was illustrated by beautiful plants in baskets and ceramic pots lining the street, even though it was cool and cloudy. It reminded Mark of the wet, cool, winter rainforest climate of the coastal Seattle area, of the northwest United States back on Earth. Mark felt the chill and adjusted the temperature of his suit using the control knob above the front pocket of his coat. He continued down the street until he found the address he had been given of a small storefront hidden among the many shops lining the street. Mark entered the store and saw a short man with dark brown hair, bushy eyebrows, a pointed nose, and a wide face wearing a black and white checkered vest fitting the description he had been given, working on an old mechanical clock behind the counter. Mark approached the counter and came right to the point. "I am looking for my client," he said.

"Who's asking?" asked the clerk, uninterested.

"Nobody!" said Mark.

"And who is your client?"

"Rhoda!"

The store clerk smiled. "I have a note for you," he said as he handed over a handwritten note.

Mark read the note, then asked, "You have the pack for me?"

The store clerk set a small backpack on top of the counter.

"How much do I owe you?"

"It's paid for," the store clerk smiled again.

"Thanks." Mark slipped the backpack over his shoulder and left the store. He continued toward the edge of town until he found the stable mentioned in the note.

Mark entered a small office adjoined with the stable and saw an android sitting behind a large wooden desk. The android looked up and said, "Bonjour!"

"Bonjour," Mark replied. "I have a reservation for two bulls." He was referring to an English term for a French Camargue adult horse originally from southern France. Camargue stallion horses, despite their small size, possessed the stamina, hardiness, alertness, and strength needed to carry grown adults and were adaptable in Francia's wet environment of marshes and wetlands.

"Your name?" the android asked.

"Fisher," Mark replied.

"Come with me. It's all ready and set up for you."

The android walked through the side office door to the stable. Two white horses that looked like nothing Mark had ever seen stood before him. The bulls' bodies were wider and somewhat shorter than a regular horse. Their hooves were much broader and fitted with saddles and saddlebags. "How are they different?" Mark asked, referring to the horses.

"Mostly, what you see is still a Camargue stallion. But our geneticists have modified their hooves to resemble a camel's wide feet because of the soft moss that grows on this planet. The horse is tailless because there are no insects to bother them. Have you ridden a bull before?" the android asked.

"Many times, long ago," Mark answered.

"Where is your partner? You have two bulls," the android asked. "My partner will join me later," Mark replied.

"That's good." The android pulled an electronic pad out of his shirt pocket. "Please sign here for the authorization."

Mark signed the name 'Fisher' on the pad, handed it back to the android, tied the halter of the second animal to the first, and led the bulls out to the street.

Mark mounted the lead animal and rode to the forest edge beyond the stable. He rode along a well-worn trail through open fields away from the town center.

He came to a large, multi-layered rock formation in the shape of a monument twice his height beside the forest entrance with a plaque describing the trail's construction during the French migration years ago.

According to the note, this is where I am to wait, Mark thought. Mark looked around the forest and open fields behind him. He took the note out and reread the instructions, then folded it and put it back in his pocket.

A sudden voice startled Mark out of his reverie, "Hello, Captain Styler!"

Mark snapped around to see Melissa, dressed in outdoor hiking clothes. She was wearing a black hooded jacket, pants, and a shirt with calf-high boots, sitting on the other horse. Mark was surprised at how Melissa mounted the bull while the animal didn't make a sound.

"It's my pleasure to meet you again." A smile appeared on her face.

Mark smiled in return and replied, "The pleasure is all mine."

"I know you have many questions for me, Captain, but I would prefer to defer them until we make camp tonight on the other side of the forest just ahead of us. Afterward, I will explain the need for the horses and the reason for our mission."

"Of course. I understand. But first, I need something in return, Melissa. Call me Mark instead of Captain Styler, please," Mark said.

"Mark it is then, and you may call me Mel if you wish."

Mark nodded, untied the horse's rein, and handed it to Mel.

After a quick "Thank you, Mark," they set off with Melissa in the lead.

Mark-caught up with Melissa and rode beside her. The trail inside the forest was just wide enough for both horses to travel abreast. Mark could sense a soft, shallow, sponge-like feeling beneath the horses' hooves on the trail path. The tree trunks, the leaves, and the umbrella-shaped treetops were a deep, dark green. The trees were spaced well apart, allowing a thick, yellow mixture of green moss to cover the forest floor. A small filter of sunlight was breaking through the thin layers of the treetops, illuminating multiple reflections of Melissa's yellow eyes upon her light blue skin.

Mark smiled and thought, *She is pretty, strange, but pretty.*

"You are not wearing your mask this time," he said, looking forward along the trail.

"I don't need a mask here. The humidity is very much like in my home world, only it is much warmer."

"What is your home world like?" Mark asked.

"It is what you would call a water-world, smaller than Earth, but without a large land mass. My world has many small and large islands spread throughout the ocean's surface. Many of the islands experience a tropical environment, much like your home world," Melissa explained.

"Your home world must be quite beautiful," Mark responded with enthusiasm, but Melissa didn't continue further describing her homeland. He sensed that bitterness tinged her description and could only assume that there must have been a lot of destruction of that beauty due to the war.

"It's another eight hours before we can rest for the night," Melissa said curtly, and Mark kept any other questions to himself.

The further they rode along the forest trail, the taller and more widely spaced the trees became, which gave the forest a feeling of openness. As the morning wore on, the two riders ventured further into the forest until the tree trunks grew to twice the width of Mark's shoulders. He noticed silver-green vines the size of his thumb hanging from certain branches, with some of them growing attached to the ground.

"What are those tree vines?" Mark asked.

Melissa replied, "The trees with the vines are ending their growth cycle. When that happens, each tree's upper branches produce a series of tendrils which grow down toward the ground. Once the tip of a vine connects with the ground, the roots develop a rhizome and a new tree starts. At that point, the parent tree and vine die. The sunlight connecting with the forest floor increases as the top growth dies back. As sunlight strikes the rhizome, it grows, and the tree growth cycle starts. Over time, the moss-covered ground consumes the decomposing trees, branches, and leaves, leaving behind a clean, revitalized forest floor."

As the day went on, Mark realized that although he had seen many small, timid creatures, there weren't any aggressive predators, even though they were needed to control the animal population.

Mel sensed Mark's thoughts and said, "If you are thinking about why I am not concerned about predators, it is because they only hunt at night, which is why we must camp on the other side of the river before it gets dark."

As the day passed, the riders came to a bed of black basalt lava rock marking the trail's end. This vast ocean of lava had flowed long ago and spread out with small islands of cinder cones. The view reminded Mark of the craters of the Moon National Monument and Preserve, a US national monument and national preserve in the Snake River Plain in central Idaho.

"Who built the trail through the lava field?" Mark asked.

"Early French settlers built the trail, but it is no longer frequently used. The predators here don't like open ground, so we should be able to pass through safely, even at night. We need to cross the river valley not far ahead; then, we can set up our tents. The local natives built a campground sheltered by a rock wall just on the other side of the river. We should be safe there."

Mark could hear the sound of moving water in the distance. *The river is not far now,* he thought.

Mark reigned in his horse on a low bluff overlooking the valley below. The river was wide but seemed shallow, as erosion from the waterway had worked through the black basaltic lava rock. They carefully descended the low valley bluff toward the river's edge.

"Let's stop here so that the horses can drink. Hold my reins while I look at the river bed, Mel."

Mark waded into the water to check on the river bed. The river bottom comprised many loose layers of small, flat basaltic rocks covered with algae. *This was slippery but not too bad for the horses,* Mark thought. *The loose stones under the horses' hoofs may frighten them more than cause them to misstep.*

Mark decided to take caution because he did not know if the horses were familiar with crossing water. He knew that all horses are skeptical about crossing water if it's not something they regularly encounter because it's hard for them to gauge depth. *It's up to me to build my horse's confidence. If I can do this properly, it will become less about the water and more about the horse's trust in me and his confidence in himself,* Mark thought.

Mark turned to Melissa and said, "We will have to walk the horses across. The river bottom is too slippery for the horses with us on their backs."

"Our campground for the night can be seen over there above the river bluff."

Mark gazed in Melissa's direction. The Sun was so low on the horizon that it was hard to see.

Mark suggested that Melissa speak to the horses before slowly leading them one at a time into the river's widest point. For most of the crossing, the river was no more than a foot or two deep, but the stones on the river bottom slid around quite a bit, and Mark was glad that he had taken the time to check the river bed.

They led the horses out of the water on the other side. The black basalt bluff fell away from the river's edge as the horses gradually made their way up above it to the top of the outcropping where the trail began. Once they were at the top of the bluff, Mark noticed a corral of flat stones piled high above his head toward the side of the trail.

"This is where we camp for the night," said Melissa.

They entered through the narrow entrance of the corral. One end of the 30 by 50 feet enclosure was divided into a smaller pen for the horses by a 4-foot-high rocky fence. Layers of logs blocked both entrances. Mark removed the logs as he led their horses through. A fire pit was in the larger corral where the tents would be set up.

"The natives put a lot of thought into designing this camp," Mark told Melissa as they set up camp. "I'm glad we still have light."

They removed the saddlebags from their mounts and set up their dome tents. Since the tents couldn't be secured to the rocky ground, Mark found enough flat stones to give weight to the inside edges of the tents and hold them down in the event of a storm. Melissa took a small heater box out of one saddlebag and set it on the ground between the two tents, where it quickly radiated the same amount of heat as an open fire pit.

Looking in the saddlebags, Mark said, "There is dry food for you and paste tubes for me."

"I'll be back in a few minutes to find silages for the horses," Melissa said.

Mark got out sleeping bags and blankets from the duffle bag strapped behind one of the saddles. After a while, Melissa appeared back in camp, dragging small umbrella-shaped plants to feed the horses.

"Where did you find those plants?" Mark asked.

"Over the crest of the bluff and on the other side another 300 feet, there is a grove of young plants similar to those in the forest across the river," Melissa replied, tossing a few stalks to the horses. "These plants have genetic structures very similar to the legumes the horses feed on. They will be safe for them to consume." Mark watched as the animals greedily munched on the saplings.

"The high walls around us will help to keep predators out, and the light from the heater box will also keep them away. Let's sit down now, and I will tell you what is ahead of us." Mark and Melissa squatted together between the tents near the heater box.

After they had eaten their meals and were settled in for the night, Mel began her story. "As I mentioned to you six months ago at our last meeting, my people have recovered the second black stone from the set of four stolen energy transmitter convertors. We have located the third nearby in the next valley."

"Why don't you just walk in and take it back?" Mark suggested.

"I wish it were that easy, but it is not," Melissa replied. "The first two escape pods ran out of fuel and crashed on other planets, and both the crafts and onboard computers were disabled on impact. Anyone could have taken the converters from the pods, and because of our policy of non-interference in other cultures, we need humans to do the leg work.”

“Why is that?” Mark asked.

“I will explain it this way, Mark. There are three levels of advanced technology. Level one is from the Stone Age to the Iron Age. Level two is the Industrial Age to the Space Age. Level three is us.”

“What is level three called?” Mark asked.

It is just called level three. I know this is difficult for you to understand, but I am bonded by our policy of non-interference toward all level one and two cultures to remain neutral, and we must remain unseen. In the past, our presence has created chaos in other worlds’ political and religious cultures. The Commonwealth authorities are highly aggressive toward us, making it necessary for us to conduct our business in secrecy.

“There is another problem, too, isn’t there?” Mark acknowledged.

“You are very perceptive, Mark,” Melissa replied, adding, “Let me elaborate. The third escape pod landed safely, and the onboard computer remained enabled. On Meloson all living creatures are comprised of copper-based oxygen hemocyanin molecules which make our blood a light blue, very much like your Earth’s horseshoe crab. You may see similarities between our species if you consider that your iron-based oxygen hemoglobin molecules make your blood red.”

As Mark considered this information, she continued, “Our rivals, the Walburgis, who have stolen the energy transmitter converters, have iron-based blood, and are the only level-three, iron-based presence in our sector. Each one of their escape pods will scan a given area for beings with copper-based blood and defend itself. In this sector, the escape pod computers will ignore you and all other iron-based blood types. The pod scanners may also defend against our level-three technologies, which is why I asked you to leave your black stone ring behind at your ship."

"So why don't the Walburgis come and take the stone?"

"Our home sector is extremely far away from the Orion Spur of the Milky Way, and the Walburgis are too far away to know about it. That's why your Commonwealth will not be in danger of invasion from the Walburgis Empire anytime soon and perhaps never will be."

Mark suddenly felt a chill coming on and adjusted his jacket's temperature control. "Don't you feel the cold?" Mark asked.

"My clothing auto-adjusts the temperature for me," Melissa replied.

Mark shifted, and glanced away from the heater box for the first time and noticed how dark it was. *I am beginning to understand why I am involved in this mission*, he thought to himself. "Where do I come into all of this?" he asked Melissa.

"The Walburgis' third escape pod landed in the heart of a native village located in the next valley just east from here. You and I are traveling there to find it. When we locate the crash site, you will continue alone. The escape pod looks like a white, slightly flattened egg about 10 feet in length. On the top of the pod, there is an oval-shaped red outline." Melissa pointed to the backpack Mark had picked up. "In your pack, you will find a sealed black glove, which would fit the shape of your right hand, Mark. The glove has also been designed to resemble the size and shape of our rival's hand, and so it is twice as big as yours, yet it will still be easy for you to maneuver and manipulate. The Walburgis' DNA coats the glove's surface. Break the seal only when you are in the escape pod, so the glove doesn't get contaminated by the local environment. Put on the glove. With your right hand touch the hatch in the middle of the red outline. It will slide open, revealing the black stone. Remove the stone from the hatch, and return to our hiding place as quickly as possible."

"This process sounds too easy," Mark said, "What's the catch?"

"The villagers have a primitive level one society. The locals live a very simple life in mud and straw huts. Our assumption is that village healers ran to a nearby field to see the strange craft in the sky. The pod killed the healer when it landed on top of him."

"Pity, but then he wasn't so smart to be in the way," Mark said.

"No, but the villagers worshiped him as if he were a god. Since the escape pod killed the healer, the villagers now worship the pod. The villagers won't let any outsiders come near it."

"Oh great!" Mark murmured. "So how am I to approach the pod?"

"In your pack, you will also find the same type of clothing the villagers wear, along with a red wig that mimics native hair. The best time to approach the escape pod is in the morning during meal time. The natives should all be in their huts then."

Mark turned his head toward Mel and asked, "And when is that?"

"About an hour after sunrise. You are a little taller than the villagers, so I suggest you hunch over a little," said Melissa, adding a warning, "There is an element of risk, as the villagers are very protective of their pod."

Mark replied with assurance. "Mel, everything I have been successful at in the past has had an element of risk involved in it. If I hadn't taken risks, I wouldn't have achieved my goals and gotten to where I am today. The unknown factor in a mission is what I worry about, and I don't see one at this point. Besides, I signed on to this assignment, and want to get on with it."

Melissa smiled at Mark without saying a word.

For an alien, she has pretty eyes, thought Mark. *I enjoy having her along. Something I have missed since Kate's passing is female companionship.*

"It is time for bed. See you in the morning, Mark," Melissa said as she rose gracefully and disappeared inside her tent.

Mark replied gently, "Seems like I am going to have an interesting day tomorrow."

Stiffly he stood up and stretched, entered his tent, and slipped into his sleeping bag. He immediately fell into a deep sleep, quite exhausted as he was.

Chapter 4

From the top of the hillside the next morning, Mark and Melissa lay on their stomachs and looked out over the village below. Wafting smoke rose lazily from hearths inside the three or four rows of mud huts arranged neatly side by side. In a field nearby, Mark spied the escape pod.

"This shouldn't be hard." Mark's spirit was picking up. "Wish me luck!" he whispered to Melissa.

Mark crept carefully down the slope, moving quietly from one tree to another until he reached the edge of the forest. Most of the villagers were inside their huts as Mark approached.

Hunched over and trying to walk with the shuffling gait of a native, Mark made his way toward the pod. As he neared it, some of the villagers noticed him; however, they accepted his presence and turned away. Mark was soon so close that he could see the red outline on top of the pod, just as Melissa had described. Mark removed the sealed package from his pack and put on the glove.

With his right hand enclosed in the glove, he carefully touched the middle of the panel. It slid open, revealing a black stone. Mark quickly removed it and placed it securely in his pouch, then casually shuffled his way back toward the forest. Melissa was at the top of a bluff behind a rock face.

That was easy, thought Mark.

Suddenly, Mark tripped, and heard a crunch and then a terrible squeal. Glancing down in disbelief, he realized that he had stepped on a small animal. The creature made a dreadful clamor as it ran in a tight circle. At the sound, the entire village emptied from their huts amidst an uproar that sounded to Mark like a flock of gobbling turkeys. Mark bolted across the field for the forest, and ran up the hillside.

The villagers ran en masse for a nearby pen and released a group of animals, which immediately made their way toward Mark. They were stocky and short with a mixture of short yellow and black fur, with a thick mane of long, tough back hair from their neck to the base of their tails. They were muscular creatures, made for the chase, with a short torso set on long, thick and powerful legs. Their bulky necks supported heavy heads with small eyes protected by drooping eyelids and sharply pointed ears set on stubby skulls, their open jaws showing gaping teeth which, to Mark, seemed perfect for a predator.

As Mark sprinted up the hillside, he could hear their grunting noises gaining on him.

Mark yanked his stunner out of his shoulder holster and fired at the pack leader. The stunner struck the beast, connected with its nervous system and it collapsed as it fell unconscious. Although his aim was good, Mark didn't like what he saw when he looked back. They were gaining on him. Soon he would be surrounded by the animals, and he could hear the natives yelling and screaming encouragement as they followed closely behind the pack.

"Mark!" Mel shouted as she dashed down the hillside, using the heavy bodies of both horses to break through the line of attack. Melissa skidded to a stop near Mark as he rushed toward his mount.

With a flying leap, Mark scrambled onto his agitated horse just as another of the pack leaders slashed and clawed at its rump. Another animal jumped high and slammed into Melissa, knocking her to the ground. She swiftly rolled to the side, kicking out at the beast as she tried to remount her wide-eyed, side-stepping horse. Mark quickly let off several shots from his stunner as he struggled to control his mount.

With the frightened horse twirling around in a tight circle, Melissa, frustrated as an inexperienced rider, somehow managed to regain her seat, and they raced back toward the river to the west. Suddenly, a villager stepped out from behind a tree armed with a loaded leather slingshot and blocked her path. A rock hit Melissa's left shin. The terrorized horse charged, trampling the native. Trying to control his stampeding mount, Mark skittered around the downed native, following as closely as possible to his injured companion.

As more rocks rained down upon the riders, they began to outdistance their attackers. The river was barely in sight when Melissa's horse stumbled on the loose pebbles as they scrambled down the bank, landing heavily on the rocks below. There was a terrible crunch. The bull screamed in agony and Mark realized that both its front legs were broken.

Upon impact, Melissa flew over the horse's head and landed with a large splash in the river. Mark brought his mount to a halt just as the pack caught up with them. With a weird howl, a huge beast bounded forward and grabbed the throat of Mark's mount. Mark was lucky enough to slide out of the saddle as both animals tumbled down the riverbank together. He quickly pulled out his stunner and shot the critter. However, he could see that he was too late. His horse was on its side by the river bed with a ripped and torn throat. Mark frantically looked around for Melissa as he removed the saddle straps from the dead animal, and with all the strength he could muster, removed the saddlebag from underneath the carcass. He spotted Melissa struggling in a shallow pool of water, unable to stand up and wade the short distance to the river bank.

Mark made his way toward Melissa, slipping and falling several times on the algae-slick pebbles before finally reaching her. Pulling her up quickly and half-carrying her to the shore, Mark could hear the gobbling noises of the villagers getting louder behind them.

"I can't stand!" Melissa cried. "My leg is bleeding!"

Mark slung Melissa over his shoulder, and they quickly made their way toward the opposite side of the river. Deadly torpedoes hurled from the natives' slingshots landed in the water around them. Finally, they reached the opposite shore. Mark carefully set Melissa down on the dry rock bed, then collapsed beside her. Exhausted, wet and cold, they looked back across the river. The villagers and their packs were still lined up along water's edge, pointing, shouting and snarling their frustration at the travelers.

"What is stopping them from following us? They seem to be afraid of something," Mark asked.

"Look over there, just a little upstream from the villagers," Melissa replied and pointed.

An assembly of black heads poked up from the deepest part of the water, glaring at the villagers. The partly submerged bodies made it hard to tell the size of the creatures, but judging from the villagers' apprehension, they were most likely treacherous.

"What are they?" Mark asked, "Were you aware of any predators in the water?"

"No, I didn't, they are new to me."

"Let's get away from here and make for higher ground," Mark suggested.

"Mark, I can't walk, my leg is badly injured."

Mark felt unprepared to administer first aid to Melissa. He knew little to nothing about her physiology. He asked Melissa, "Could you instruct me on how to help you? You have the technology to assist me with that don't you?"

"Without my first aid kit, there isn't much we could do for my leg." Melissa agreed that if they could move to a safe place, she could instruct Mark about what to do.

"Then I'll have to carry you back to town," Mark decided.

"No, this mission is too important for my people. Leave me here and use your ring back at your ship to contact my superiors," said Melissa.

"I don't care how important this mission is; I'm not leaving you here," Mark exclaimed.

The black creatures were approaching them at a sluggish pace. Mark reached for his stunner on instinct but the holster was empty. *The stunner must have dropped into the river!* Mark realized. Quickly and without saying another word, Mark placed his saddlebag over his shoulder, checked to make sure the black stone was still secured in his waist pouch and heaved Melissa into his arms. Shushing her grumbles of surprise, he half-marched, half-stomped away from the river until he felt reasonably sure they were safe.

As they were off the beaten trail, the lava rocks of the path were uneven and had sharp edges. Mark grunted as he struggled with trying not to fall over the rocks. The Sun was almost directly overhead and with a great sigh he muttered, "I have to rest." He carefully set Melissa down on a small boulder, then dropped down to the ground beside her. After taking a few minutes to catch his breath, Mark glanced around and realized that nothing looked familiar. *The trail should be further north,* he thought to himself and realized that they were quite lost.

"Let's look at your leg," he said, lifting himself to a sitting position.

Grimacing and obviously in pain, Melissa answered "Forget my leg, just go on without me. Mark, please, take the stone and go!"

"Get it through your head, Mel, I am not leaving you!" Mark fumed.

"Then I will have to shut down my heart," said Melissa.

Mark, not sure what Melissa meant, gently took hold of her face, tilting her chin so that he was looking into her beautiful eyes.

"Now you listen to me," Mark said to Melissa, "Since my wife, Kate, died, I have made a vow to myself that I would never allow anyone to die on me again! If you shut down your heart, I'll throw this black stone into the river! Got it?"

Melissa looked into Mark's eyes as if she was reading his mind and said in a low voice, "Got it!"

Mark took hold of the blue fabric at her ankle and rolled up her pant leg. There was a nasty blue cut on the front of her shin. Mark's clinical training took over as he felt for broken bones, but he couldn't pinpoint either a fracture or a break.

"Other than a nasty bruise and a cut, it seems like your leg isn't broken, but there may be a simple fracture that I can't feel. What is your opinion Mel, can you feel anything when I touch your leg?"

"Your prognosis is the same as mine," Melissa replied.

"Will you be alright for a few minutes?" Mark asked. "I need to get out of these wet, freezing clothes. Mark stripped down in front of Melissa and reached for his saddlebag. Gasping, he stopped - one of the saddlebag pouches was torn open, and some of the contents were missing! Mark was in a near panic, not realizing he had been hauling them the whole way. Luckily, his shirt and pants were still there, but his temperature control jacket was missing. *Great,* he thought sardonically. *One more thing to go wrong.* Quickly checking the other pouch, he was relieved to find that the second pouch was intact; one blanket and the dry food were still there.

"You're naked!" said Melissa mockingly.

Here we are in the middle of a dilemma, and she is either teasing me or something else. Why? What's the point of it? Mark rolled his eyes and dumped the blanket over her head. "Keep it there until I get dressed!" Mark demanded. Mark muttered to himself; *that darn alien knows too much about men's modesty.*

Mark quickly dressed. "Okay you can uncover your head now," he said.

Melissa removed the blanket from her head and wrapped it around her body. "Let's make a plan for getting back to town," Melissa suggested. "Do you have any ideas on how we are to get back to town?"

"I'm working on it," Mark replied, knowing he hadn't worked out the remainder of the trek yet.

Melissa ignored Mark and continued, "I am too heavy for you to carry back to town. You will tire quickly after the immense amount of energy you have expended so far. You are better off leaving me behind. You need to understand our predicament and the predators that wait for us in the forest."

Mark, tired and confused by Melissa's sudden personality change, cut her off by glaring at her with his pointed forefinger.

Melissa smirked and sarcastically replied, "Got it!"

Mark glanced at his watch and said, "We have another six hours of sunlight left, so we've got lots of time before it gets dark to reach the forest."

"You might as well eat something, then," Melissa said. "You are going to need as much energy as you can gather to carry me back to town and to help you think more clearly." She continued with a worried look, "What if you injure yourself carrying me? What will we do then, Mark? You will fail our mission."

"I told you, I am working on it!" Mark felt the heat rising in his face.

Melissa smiled, hugged her one good leg, and tilted her head toward Mark. This was not the reaction she had expected from Mark. *The more I try to aggravate Mark, the more stubborn he gets. Human psychology is more complex than I expected. I have to give Mark credit though for his determination to succeed in this mission, even though if he helps me it will fail. I must think of another tactic.*

Darn it, Mark thought, *she seems too human when she looks at me that way. All it does is make it harder to leave her behind.* He began to feel more helpless as he understood that Melissa was right about the necessity of leaving her behind.

"Lord, I need your help now!" Mark muttered to himself.

Mark picked up his saddlebag and was about to throw it away. Then a thought came to him. Mark looked at Melissa, then back at the saddlebag. *Why not? It might work. I think the Lord just answered my prayer.*

"Did you think of something?" Melissa asked.

"Yes, I am going to make a yoke," Mark smiled.

A puzzled look appeared on Melissa's face.

"Bear with me, Mel," Mark insisted.

He took a knife out of his pouch and cut a large hole in both sides of each saddlebag and buckled each saddle's cross strap together. Satisfied, Mark lifted Melissa and slipped each of her legs into a saddlebag hole up to her knees. Mark loaded the remaining food into each pocket and a water jug into his waist pouch. Facing away from Melissa, he put his head through the central hole of the saddlebag cross straps and stood up. Melissa's arms circled his chest, with her legs arched forward and her shin angled down.

"I think this yoke is going to work! You are surprisingly light!" Mark said, feeling pleased with himself.

"I underestimated your ingenuity, Mark," Melissa replied. "Congratulations!"

"Thanks! Are you alright back there?"

"Yes, very much so!" Melissa said, although she was still grimacing.

"Then we're off! Do you know where the trail is, Mel?"

"I believe it's over to the right!"

With new enthusiasm, Mark quickly found the trail as it turned left. "I guess we will reach the forest in about an hour."

"At this walking rate and with several short rest periods, 1.6 hours," Melissa corrected.

"Add another four hours inside the forest, and we will be spending tonight in the middle of it," Mark realized.

Melissa interrupted his thoughts. "Have you thought about night predators?"

"I know, I am working on it," Mark replied.

There doesn't seem to be much wildlife among the black basaltic lava, Mark thought. *Perhaps the predators are too efficient? The predators were probably responsible for the lack of wildlife. And I haven't got my stunner, heat box, or jacket. Just a lousy small camping knife and one blanket,* Mark thought wryly.

He was sweating and feeling exhausted again when the forest edge came into view. They had left the western side of the lava flow behind. Mark looked at his watch. *Wow, 1.6 hours, just like Mel said it would take,* he thought.

"Let's rest here and eat before entering the forest," Mark said, lowering Melissa carefully onto a nearby rock conveniently set waist-high. He slipped out of the leather yoke, then took the last tube of food paste out of his pocket and gave it to Melissa, who was studying his every move. When Mark had finished the last of his food, he removed the cap from the water canister and took a few sips, then passed it to Melissa.

As she drank, he asked, "How does your leg feel?"

"It's still sore, but I think I should be able to walk in a few days," Melissa replied. As Mark checked the leather yoke for wear, she added, "Have you come up with a solution regarding the night predators?"

Irritated and tired, Mark took a breath to avoid exploding. "Look, Mel, I said..."

"I know, you are working on it," said Melissa.

Then it hit him. Melissa was still trying to provoke him to manipulate him into leaving her behind. Mark took another deep breath to regain a sense of perspective about the situation.

Melissa smiled at him and said, "I am ready to continue if you are."

Mark, nodding, slipped his head through the yoke and entered the forest without saying another word.

The heavy branches overhead shielded them from the hot Sun and helped to dissipate any random breezes through the meadow. Mark felt his body slowly adjust to the forest chill. Time seemed to pass slowly as the travelers moved in a dream-like trance along the forest path. Mark asked, "How long have we been walking, Mel?"

"About 3 and a half hours," Melissa replied.

"3 and a half hours," repeated Mark, suddenly feeling weary and with renewed concern about the coming nightfall. "What do you know about the predators?" he asked.

"They are similar to the pack the natives called up to attacked us earlier," Melissa replied. The French settlers call them bush wolves, and the night predators are called forest wolves. They are hairy animals like those that attacked us, with green, yellowish fur, speckled dark gray to black. The heads and bodies of the forest wolves are longer and narrower than bush wolves to suit the cool and wet environment of the canopied forest in which they hunt. The forest wolves have shorter front limbs than the bush wolves who have long hind legs that can launch them up to 20 feet in a single leap! Their extra-large paws help keep them from sinking into the thick forest moss. Also, they have one major advantage over the bush wolves."

"What?" Mark asked.

"They can climb trees," said Melissa.

Oh boy, Mark thought, *they can even climb trees! This is not going very well; there is not even a stick nearby in the forest that can be made into a spear to protect us. There was no underbrush, just a moss-covered clearing between the umbrella trees. The mosses did an excellent job dissolving the dead branches and trees. I'm exhausted. I had better stop here and get some rest and clear my mind to think things out more clearly.*

"I am tired. Let's rest here for a minute. Can you use the weight of your good leg and lean against this tree?" Mark asked. "It would be easier for me to slip off the yoke that way."

"I should be able to." Melissa was feeling tired too.

First, Melissa moved, then Mark bent down to his knees, slipped Melissa out of the yoke and sat down with his back against a tree, eyes closed. Melissa watched Mark as he muttered to himself wordlessly.

"What are you saying?" Melissa asked.

"I have just stepped out of one yoke, and now I am asking for a yoke of another kind."

Puzzled by Mark's riddle, Melissa sat beside him, "Are you meditating?" she asked.

"Yes. That is what I do when I am feeling depressed, asking for directions from God or giving thanks," Mark replied. "Allow me a few minutes, Mel."

When Mark opened his eyes, he saw an amazing sight. A beam of late-evening sunlight created multiple reflections amongst the hundreds of glistening vines hanging from the overhead umbrella trees. Mark moved among the glowing vines, gazing at the wondrous view above him. He smiled at the wonder of the vista before him. The answer to their ordeal suddenly became clear to him at that moment.

Mark looked among the trees for the most condensed group of vines. He took out his knife and cut the lower portion of the vines from the ground. Mark grabbed the cut vine and used his weight to test its strength, "It holds!" Mark cried.

Mark picked Melissa up and carefully lowered her down until she was sitting beside the heap of cut vines. Then he explained his idea: "I am going to climb this main central vine, and when I point to another vine, bring it to me, and I will take hold of it."

Mark climbed 25 feet up, made a double fireman's loop, and slid his leg through each of the two loops to support himself.

"Mel, bring those other vines beside you to me one at a time!" Mark pointed with his hand as Melissa struggled to pass up the vines to him one at a time. Mark took each vine from Melissa and made a fixed loop as the remaining vines hung down. After he had finished with the other nine vines, he grasped the vines hanging below the loop and tied them to the opposite loops. Mark let the remainder of the vine hang down. Melissa caught on to this technique quickly and arranged the next vine. Mark asked Melissa for more vines as he cut and weaved from the center of the base in a circular process until the basket platform was six feet across.

Mark smiled, bent over the edge, and said to Melissa, "Now we have a hanging platform!" But just as quickly, Mark's smile vanished. Two giant forest wolves were creeping up slowly behind Melissa.

Mark shouted, "Melissa! Quick, climb the vine now!"

Melissa, sensing danger, somehow managed to pull herself and her tender leg up the vine just as the wolves rushed toward her. Melissa was surprisingly quick, but before Mark could reach for her, the wolves had seized the main vine's end and violently shook and pulled down on it. Mark desperately struggled to get Melissa, who dangled precariously in the air only arm's length from Mark.

"Give me your hand!" Mark cried.

Melissa's hand grabbed Mark's as her remaining strength gave out, and she let go of the vine. The forest wolves were wildly trying to jump at Melissa, who was dangling from Mark's grip. Mark used all his strength and, with both hands, lifted Melissa up, over, and onto the tangle of vines that made up the platform. At that moment, a final tug from the predators on the central vine shook Mark, causing him to fall from the platform. His fall was cushioned by one of the creatures' back as he rolled to the ground. The startled forest wolf leaped up into the air and bumped into the other animal, which let out a primordial, high-pitched scream, and attacked the wolf that had bumped into him. Combat broke out between the two beasts.

Taking advantage of the wolves fighting each other, Mark quickly grabbed the saddlebag and blanket, threw them over his shoulder and hurried back up the central vine to the platform. Melissa struggled to help Mark back onto the platform, and they both lay on their backs, breathing heavily.

Mark gasped, "I thought I had lost you."

"I think your God made sure I was okay," Melissa said, panting.

"I'm going to pull the center vine up. That will keep the wolves from shaking us off the platform again."

Realizing their prey was beyond their reach, the two night predators circled the platform and eventually lay down close by, staring up at Mark and Melissa with hungry, glaring eyes.

Mark, hot and sweating profusely, whispered. "I think we can rest now for the night."

Melissa pulled herself beside Mark and wrapped the blanket around her body.

With darkness almost upon them, Mark asked, "You have been edgy for the last three hours. Would you like to tell me what is on your mind?"

Melissa looked at Mark, hesitated for a moment, and then said, "I didn't think we would make it this far into the forest and also be able to secure ourselves for the night. Mark, I want to apologize to you for trying to provoke you into leaving me behind. Your determination and dedication in completing this mission remind me of someone very close to me."

"Is it your partner back on your ship?"

"No, he died in our home world during the war, which was a long time ago." The painful memory caused her to turn her head away. She closed her eyes and said, "Good night, Mark."

Mark didn't know what to say or how to react to her painful memory. The platform gently swayed about as Mark carefully stretched out on his side and whispered, "Good night."

***

During the night, the temperature gradually dropped until it settled to near freezing. A sudden trembling of the platform awakened Melissa. Alarmed but only half awake, she scanned the forest below for danger, then realized it was coming from Mark, who was shaking uncontrollably from the cold.

Melissa knew she would lose Mark by morning if she didn't do something immediately. Could she dare do what it would take to save him?

Melissa inched closer to him until she could feel the ridges in his spine, then placed the blanket over them both and cradled him tightly. Melissa should have felt repulsed by this closeness to an alien, but she wasn't troubled by it, as she felt a strangely natural reaction to this member of an interracial off-world species. But then Melissa remembered how she wasn't bothered when he carried her. *How strange,* Melissa thought. Mark felt a slight movement and opened his eyes. He was surprised to feel Melissa's closeness as she lay sleeping while holding him tightly. *Oh, what the heck,* he thought, and closed his eyes again. Gradually, Mark's shaking subsided, and Melissa slept.

***

The warmth of the Sun on his face caused Mark to wake up. The thick fog and dew on the ground were being heated by the Sun and evaporating like steam toward the sky. Mark pulled back the blanket and turned to see Melissa sitting up and holding her good right knee tight to her chest.

"Good morning. You are looking better this morning," said Melissa.

"Thanks. How about you? How is your leg today?"

"The feeling is starting to come back to my thigh, so I know the healing process is moving forward," Melissa continued, "I had the most interesting night! You were talking in your sleep."

"Why is that so interesting?" Mark asked.

Melissa ignored why and asked, "Who is Katherine or Kate? After you had settled down during the night, you mentioned her name in your sleep. Is she your wife?"

Mark moved closer to Melissa and placed the blanket over their laps.

"Katherine," replied Mark, "was my wife. She died about five years ago."

"I would like to hear about your Katherine," Melissa asked.

It was like a wall had been broken between them, as for the next several hours, they shared stories of their lives, their home worlds, their dreams and painful memories.

"The ground is dry and there's a bright daylight," Melissa said. "The forest wolves will be sleeping in their dens for the day. It is time for us to continue with our journey."

They gathered their few possessions. Mark lowered the central vine from the platform to the ground. While Mark pulled back on the vine for extra support, Melissa dropped the saddlebag and blanket to the ground and climbed down. Mark followed her and landed on the ground. Melissa was leaning against the tree, ready with the yoke, which she helped slip over Mark's head.

"How many hours until we are out of the forest?" Mark asked.

"Three hours," Melissa replied.

For the rest of the journey back through the forest, Melissa was quiet except for the occasional "yes" or "no". They were both relieved, and some of the tension of the journey left their faces as they moved beyond the confines of the forest, and eventually, they could see the town in the distance. The old town of Francia, with its two-story ironwork balcony, was overshadowed by the modern city just beyond, bringing a welcome sigh of relief from Mark and Melissa.

Melissa suddenly cried out, startling Mark and causing him to spin around and nearly stumble. "Are you alright?" Mark exclaimed. "How is your leg?"

"I'm sorry," Melissa replied. "But my people are not nearby. Something must have caused them to leave the area. I think our security system has been breached. I not able to sense any of our agents in the area."

"What about the operator at the store?" Mark asked.

"I have never met the agent in person or the android at the stable. We use different security agents to cover our tracks; otherwise, the Commonwealth authorities would notice our appearance, detain us and impede our mission. We must return to your ship now. I will use your ring at your ship to contact my people."

Mark felt weary, "We will just have to keep you out of sight. I think starting at the stable is safe. We have to compensate them for the horses anyway. Then we'll find a way to get back to my ship. Let's get going."

***

It was well past midnight. The small spaceport terminal

was quiet and empty of human and native passengers. Boris, the security agent, was alone and half asleep, sitting on his chair with his feet up on the desk.

It's another boring night, but it's just the way I like it, he thought. At the same moment, a small alarm at the main entrance gate beeped as a man pushing a wheelchair appeared. A caregiver was transporting someone in a protective decontamination suit to a new medical facility. Annoyed that his snooze was interrupted, Boris reluctantly stood up to inspect the situation.

"Name?" he inquired of the man behind the wheelchair.

"Captain Mark Styler of the *Marvel*," the man said.

Boris held a scanner to Mark's eyes, "Okay, you can pass. Who have you got here?" he said, pointing to the silent figure in the wheelchair, wearing sunglasses inside a protective suit and an internal breathing apparatus.

"A patient that I'm taking to a specialist on our next planetary stop," Mark replied.

Boris didn't like the blue skin he saw on her face. "She doesn't look very good," he observed, studying her carefully.

"Here's her ID," Mark added quickly. Boris scanned the disc in Mark's hand.

"How can I let this pass? I can't scan her eyes." Boris grumbled.

"Do you want me to break the decontamination suit seal so you can?" Mark asked.

Boris realized that having a decontamination patient in a security room without a doctor or nurse present could be asking for trouble. And to scan her eyes? Was it worth the risk of infection? "Ah, forget it. Just get her out of here," he said grudgingly.

"Thank you!" Mark gratefully replied as he pushed the wheelchair quickly past the security gate and out the exit door.

Once outside, he made for the *Marvel* across the open pavement. Bill was waiting for them at the bottom of the ramp.

"Hi, Mark," Bill said, "Good to see you back!" as he helped Mark lift the wheelchair to the ramp. "You had us worried for a while."

"Glad to be back, Bill! You and Mason did a good job providing the medical equipment and the hotel arrangements for us the past two days," Mark replied.

"How is Melissa's leg?" Bill asked.

"Her leg is still tender, but she can walk," Mark replied. "Lock down everything immediately," he said, moving quickly onto the bridge. "Sara!"

"Yes, Mark?"

"Have Chan and Mason ready the ship for off-world. I will give you new orders as soon as Melissa contacts her people."

Mark helped Melissa out of the wheelchair and assisted her to her quarters.

Mark positioned Melissa on the edge of the bed and said, "I will get the ring from my quarters while you get out of the medical suit."

By the time Mark returned with the ring, Melissa was out of the suit and sitting on the side of the bed. She took the ring from his hand and closed her eyes, linking the ring with her mind.

Mark watched Melissa communicate with the ring and thought it must be similar to how he was linked with his implant.

At last, Melissa raised her head and said, "It looks like I was right about the security breach. The Commonwealth agents are hiding out among the planets in this solar system. That is why my people left their posts. I have related our adventures to my superiors."

Mark, relieved that Melissa could contact her people, asked, "We are almost ready to lift off-world. Do you have the set of coordinates for us?" Mark asked.

"Yes, I have them," Melissa replied.

"Do you want the wheelchair?" Mark asked.

"Don't bother," Melissa said. "If possible, I'll lean on you to walk to the bridge."

***

"Warp jump's out 5, 4, 3, 2, 1," said Chantal. "We're back in ordinary space, and the scanner shows we have a visitor. I am unable to identify the new ship's identity."

"It's alright, Chan. It's one of the Meloson scout ships," said Melissa.

"Mel, how do we get you aboard your ship?" Mark asked.

"Align the *Marvel* within twenty-five feet of the exit portal and parallel with the red outline of the scout ship," Melissa replied.

"Do you understand the instructions and copy, Chan?"

"Acknowledged," Chantal said.

Mark got up from his chair and moved to sit beside Melissa.

"I'll walk you back to the exit," Mark said.

"I would appreciate that!" Melissa smiled and added, "Put your arm around my shoulder for support."

A tube extended from the Meloson ship and attached to the *Marvel's* exit door.

"If and when your people find the last stone," Mark began, "Will you be able to contact me again?"

"Mark, your service will be most valuable, and I will try my best to mention your name to my superiors." She paused then looked into his eyes, "I will miss our acquaintance," Melissa said.

"So will I," Mark replied. "So will I."

"Where will your people go when you find the last stone?"

"I'm sorry, Mark, my superiors will not allow me to give out that information," Melissa replied.

"Then I guess this is where we say goodbye." Mark touched the scanner, opening the exit door.

Anticipating an empty portal, Mark was surprised to see a single Meloson standing before them as the door slid open.

"Hello, Captain Styler. I am here to assist Melissa," the Meloson announced.

"Take this pouch then. The stone is inside it." Mark responded.

Gratefully, the Meloson took the pouch and helped Melissa into the exit portal. As Melissa stepped inside, she turned and bowed low to Mark. The door slid noiselessly shut. Mark remembered from their conversations on the forest platform that a farewell bow is the Meloson's highest form of respect to another.

Mark closed the *Marvel's* exit portal and returned to the bridge.

"The Meloson has just warp jumped," Chantal announced.

"Chan, systematize for our next destination." Mark command.

Memories flooded Mark's mind as he bid goodbye to Melissa. She had a habit of folding her hands together just like Kate did. She had bowed exactly the way Kate used to during their first meetings. When Melissa tilted her head sideways and smiled, all Mark could how Kate did the same all those years ago. They way Melissa hugged her knees to her chest and silently watched him, he remembered Kate doing the same at the palace bench on Planet New Sweoland or watching him read a book on *Marvel*. How Melissa had held his arm on the platform in the forest to keep him warm, Kate had done that very often.

"Mel, you are so much like Kate," muttered Mark sadly. "I am still trying to get over losing her!"

"All personnel are secure," said Sara, pulling Mark out of his memories.

Warp jump begins in 5, 4, 3, 2, 1.

Chapter 5

Chantal leaned back in her pilot chair with a fresh cup of coffee. *Four days' travel in warp space is starting to give me cabin fever.* Chantal thought, *this night shift was going to be a long one.*

"Hello, Chan. Would you like some company?" Sara asked.

"Sure," Chantal replied. "Sara, my one-year contract is almost up, and I still don't know if I should apply for renewal."

"Is that why you have been so restless lately?" Sara inquired.

"It shows, doesn't it?" Chantal yawned. "How long is the layover on our next stop?"

"One day," said Sara.

"Why is the layover so short?" Chantal asked with surprise.

"It is not a good place for a layover," Sara replied, continuing, "An ultra-conservative religious order rules the next planet. They are part of the reason for the Second Great Departure from Earth. Their governing clerks have a dim view of androids. Mark doesn't think much of their radical viewpoints. If you are going off-ship, have someone with you."

Chantal's sense of rationality took over, and she replied, "I can take care of myself. I need to get off this ship so I can clear my mind. Then I can decide what to do about my contract!"

"Good evening, everyone!" Mark said as he sat down. "Report, Sara!"

"Everything is in order, and off-warp is 40 minutes away."

"Chan, you might as well take a break off-ship after we land. I would like you to give Tim a hand in the supply store. We are short on spare parts and need to purchase some extras."

"Sara, did you give Tim the list of parts?" Mark asked.

"Yes, but the supplier is on the other side of town. Do you think it is safe for Chan to leave the ship?"

"My former employee, Andy, was an android, and he didn't have any problems, so it should be okay for Chan," Mark replied then turned to Chantal. "Chan, has Sara told you about the Kazakhstani religious order?" Mark asked.

Kazakhstan! So that's who they are, Chantal thought. "Yes, Sara told me, and I will gladly stretch my legs." Chantal was now even more firmly determined to get off the ship.

***

Tim and Chantal carried their shoulder bags as they walked through the town's active corridor.

Tim said to Chantal, "It is not much further. I hope you don't mind all this walking."

"I enjoy a good walk, Tim," Chantal replied.

95

"Things have changed since the last time I was here, though," said Tim.

"What changes do you see?" Chantal asked.

"Some of the stores are closed. The streets are not as clean as the last time I was here. See that open food market over there? None of the vendors are smiling. Something is not right; they are either staring or pointing their fingers at us."

Or at me, Chantal thought.

"Here we are," said Tim.

Chantal glanced up above the door entrance. *Stan's Electronics* read the faded lettering. However, a new name had been painted over the original sign— *Kazakhstan Electronics.*

Chantal followed closely behind Tim inside the store. The building was small but packed with shelves of electronic components of every description.

"You!" a manager said, pointing at Chantal. "Sit over by the window! There will be no defilements in my store!"

Tim looked at Chantal with an expression of shock on his face.

"It is alright, Tim. I will be waiting by the window."

Chantal sat down respectfully on one of the two chairs before the window. There was a small table placed between them. She noticed two black Bibles on the table. Chantal watched as Tim handed the manager behind the counter the list Sara had provided.

"Give me a few minutes to gather these parts for you," said the manager. "I will be right back."

A junior clerk stepped into the room, and Tim stepped aside and talked privately with him. Chantal watched while the two men spoke with the manager, who moved the black book on the counter to make room for Tim's needed components.

"I need at least two shoulder bags," said the manager.

"Chan has the other bag, sir. Chan!" Tim called.

"No!" The manager shouted. "Your android can stay there!" The manager stopped for a moment and turned to Tim. "Son, things have changed lately with our leaders, and I must be careful how I do business here. Do you understand me?"

"Yes sir," Tim replied, thinking, *I'm unsure if I do.*

Tim walked over slowly to Chantal. "This is all very strange. What is going on here? May I have your shoulder bag?"

Without a word, Chantal handed her shoulder bag to Tim. When the manager had finished filling the order, he said to Tim, "That will be 200.4 credits."

"Sir, that's twice as much as the last time I was here a year ago," said Tim.

"It's all about keeping my business open," said the clerk.

Tim paid the clerk and handed the shoulder bag to Chantal, keeping the other for himself. Chantal slipped the bag strap over her right shoulder and quickly exited the store. Tim hurried to keep up with her pace.

Tim said to Chantal, "Geez, Chan, I'm sorry about what happened inside the store. I hope you weren't offended by the manager's outburst."

Still aggravated by the manager, Chantal said, "Tim, this is nothing new for me, so just forget it and let's get back to the ship."

As Chantal and Tim walked back to the ship, Chantal noticed that along the street side cafe, more black books, similar to the ones in the supply shop, were on each table. The customers either had deep conversations with each other or were absorbed in reading the black books.

"Did you learn anything from that junior clerk inside the store?" Chantal asked.

Tim replied, "There is a power struggle between two main political and religious parties. The opposition party feels that Kazakhstani society is becoming soft and morally corrupted by off-world influences. In contrast, the opposition party believes that society is moving away from the original doctrine dominated by religious clerics and revolutionary ancestors and is integrating profane constitutional changes. The economy has deteriorated because of a lack of confidence in the business sector and rising taxes.

The general public is getting angry, and tensions are escalating between the two political parties. It is becoming more of a problem for off-world traders to do business here."

"And for androids too," Chantal added, as the religious clerk seemed to consider the presence of an android as blasphemous.

"Look," Tim pointed, "there is a rally ahead."

Off to one side of the street, hundreds of local townsfolk packed into the town courtyard. The rally seemed unruly, as the mob's mood was stirred up by a man on a low platform at the end of the courtyard. A heavily built man with a dark beard held a black book over his head while crowds of people gathered around him. "My fellow believers, there is no hope for you unless you restore the path of our forefathers and reject the defilements around you!"

Chantal could see the security police moving discreetly among the people, their attention fixed on the man on the platform.

"Let's get out of here," she whispered. "Trouble is brewing."

"And there is one of them!" said the man on the platform as he screamed, pointing at Chantal with his black book. Chantal realized she was now the center of the crowd's attention and was about to be made an example of mob justice. The mob turned in unison with outraged loathing as if on cue, and one man surged toward Chantal. At that exact moment, the security police made their move.

***

Chantal, sore, bruised, and exhausted, collapsed onto her bed. Her sleeve had been torn off, revealing her bruises.

It's a miracle that I managed to escape from the police and that mob. Fortunately for Tim, he made it back to the ship untouched, she thought. Chantal removed her tattered clothing and applied healing lotion to the worst of her bruises. She struggled with her sore right ankle, limped onto the bridge, and called, "Sara!"

"Hello, Chan. Did you enjoy your time off-ship?"

"I certainly did not enjoy it!" Chantal said, annoyance in her voice. I want you to proceed with the contract discharge on the next available off-world destination."

"What happened?" Sara asked. Chantal sat down and told Sara the details of her misadventure off-ship.

"I am sorry for what happened out there, but what does this have to do with Mark?" said Sara.

"It has been almost 60 years since I left home, and nothing has changed." Chantal then realized that she had revealed too much about herself and continued to sidestep the discussion, "It's that pathetic black book, and Mark has that same book. Mark has the same viewpoint toward me as that mob!"

"Oh, Chan. Haven't you spoken to Mark yet about his Bible and the story surrounding Nana?"

"Who is Nana?" Chantal fumed.

"Mark will soon take this ship off-world." Sara added, "When Mark is off-duty, maybe you should ask him! Afterward, come back, and we will discuss the contract."

"I don't think it will do any good to have a conversation with Mark," Chantal said defiantly.

"Just do the one thing for me, and I promise you won't be sorry," Sara begged.

Chantal, emotionally drained, heaved a huge sigh. "Alright, I'll wait until after the first jump."

"Thank you, Chan," Sara replied, sounding relieved.

Chantal idly wondered why the computer sounded so emotional. *Strange,* she thought and then began configuring the jump.

***

Chantal yawned and linked her arms over her head. *My shift work is finally over, and there are only two more days of warp space. It has been the most extended trip with this crew in warp space so far.* Chantal stretched her legs and sighed with pleasure as the tension in her muscles eased. Chantal remembered that she still needed to ask Mark to tell her the story of the black book and Nana. *I wonder who Nana is,* Chantal mused. Chantal left the bridge and touched the com-link by Mark's quarters. "Captain Mark, permission to enter!"

"Come in!" said Mark from behind his desk.

Chantal entered the room and sighed when she saw the book in Mark's hand. *I might as well start here and now,* Chantal said to herself. "I have several questions on my mind lately. Do you have time to discuss them now?"

"Alright, Chan," Mark replied, putting aside his Bible.

"My first question is, what do you get out of that Bible of yours? I don't mean about giving you comfort or peace."

"Fair enough. I will give you an example of what I have read and dwelt on. I am currently studying 1 Corinthians 1:20 to 24. There are many people in this world who seek their natural wisdom. Many people have different ideas about spiritual wisdom. For example, some think that natural wisdom is the knowledge, understanding, and application of spiritual wisdom in their lives. Others consider that natural wisdom is spiritual wisdom in flowing spiritual words." he said and paused to take a breath then continued. "I like the simplicity of understanding spiritual wisdom. Like a child before his father, to submit to God is to put aside my will. I should submit to God's will and speak not from the lips but the heart. To me, that is spiritual wisdom! In this way, the willingness and sincerity of my heart could be made right. With a sincere heart, my life can be more fruitful in spirit. So you see, Chan, this is what I hoped to get out of this Bible."

Chantal grabbed a nearby chair and sat down, looking at Mark for a minute to absorb what he had said. It certainly was not what she had expected to hear from Mark. *I was expecting a rightist attitude or remarks from him,* she thought.

"I have one more question, Mark, regarding rule #27 of my current contract. 'All employees, including androids on the ship *Marvel*, are to conduct their activities professionally and respectably toward each other.' Why did you include androids in the contract? Respect toward androids is unheard of in contracts as we are considered not bona fide."

Mark smiled and said, "I wondered why it took you so long to question that aspect of the contract. The reason for rule #27 goes back to when I was 18. But it is best I go back to the very beginning. Let me explain." he smiled as he delved into his past, "My father, Eric, was in his late eighties when his first wife died. Dad married my mother a few years later, and I was born two years after his second marriage. Dad loved my mother very much. They traveled everywhere on Earth, and because my parents were away so often, I had an android who was my nanny. I called her Nana." Mark's smile widened as he remembered Nana, "When I was four years old, my mother died in a flitter accident as she was returning from an orbiting hotel. My father and I were devastated. He was never the same after that. After my mother died, Dad became introverted for a very long time. He wouldn't spend time with me or speak to anyone else."

"Dad was a co-owner and CEO of the Styler Corporation. The Corp, which is my nickname for it, was a major space drive and navigation system corporation for the Earth's Space Navy at the time. Ironically, after my mother died, Dad never left Earth because he was afraid of being off-world. In the aftermath of my mother's demise, Nana became my maternal mother. As you know, androids have very few, if any, human citizenship rights, even though they are part human. When my mother died, Nana asked my father for a program called 'Motherhood', which is an illegal download. Due to his mental condition at the time, Dad agreed to Nana's request. As a result of the program download, Nana could portray traits of affection, love, and protection similar to what a biological parent expresses to their child." Mark took a long sigh before moving on, "Dad owned a ranch just outside of Seattle, Washington. I had a private tutor and plenty of toys, but very few friends. On my 10th birthday, Dad got me a young black stallion. I loved that horse, and Nana taught me how to ride it. I rode my horse every day on our ranch. Whenever I needed her, there was my Nana; she was, in every sense, my mother and my best friend. Dad was hardly ever around our home because he also owned a penthouse in downtown Seattle. Once in a while, Dad would bring a VIP guest home to the ranch. At that time, androids were not paid as bonded servants or free agents as they are today. My father's guests often mistreated Nana, criticizing or talking down to her. I loathed to be around them, so I thought of many different reasons for Nana to take me somewhere else whenever Dad brought someone home for the weekend." Mark's smile faded slightly as his story progressed. "The year I turned 16, Dad enrolled me in a nearby private high school, hoping I would fit in or at least not be unpopular. The day I turned 18, I made new friends, helped organize the student council, and gained many leadership skills. My graduation day was life-changing in terms of my decision-making skills. My father wasn't present because he was away on business, but Nana was, and she was so proud of me. I earned an award as I was among the highest-awarded honor roll students. The school also awarded me the citizenship prize and a commendation for being at the top of the class in physics. After the ceremony, things started to go wrong when Bryson, a bully, and his three buddies decided that some things are worse than being unpopular in high school, like roughing me up because they felt I was too good for them. I tried to avoid them, but they cornered me when everyone was preoccupied with their family. One grabbed my plaque and broke it, making it look like an accident. Nana saw it happen and intervened by slapping him across the face. The bullies and everyone nearby were shocked by her actions. Even today, that has never again been known to happen. The Motherhood Program somehow overrode the android's non-interference program." A frown was now visible on Mark's face, "Someone called the authorities, and Dad, to protect himself from the authorities, wouldn't or couldn't help Nana because she illegally downloaded the program. When the Android Program Division learned about the illegal program, they sentenced Nana to the usual punishment order - a 'mind wipe'. The Android Division erased Nana's memories, and only her basic programs were restored." Mark sighed deeply. "On the last day before the scheduled mind wipe, I could visit Nana in her cell. I was in tears, knowing that she was going to be sent away and her memory of me would be gone forever. The last thing Nana and I spoke about was how important it was for me to remember the special code number for my mother's hidden safe. Then we hugged each other, and I never saw her again. Later that day, I opened Mom's safe and found a Bible and a note. The note was from Nana, explaining that the Bible was originally my mother's and how she had asked my father to give it to me at my graduation ceremony. I have kept it all these years, but I never opened it until after my wife, Katherine, died. From that day forward, I made myself two promises; one was to leave Earth forever, and the second was never to allow anyone to mistreat an android. I hope I have explained why rule #27 exists on this ship. Has that answered your question, Chantal?"

Chantal rose stiffly from her chair, strolled to the door and said in a low voice, "Goodnight, captain. I will see you later during the next shift." She walked thoughtfully to her quarters and relaxed into a chair beside her desk. She contemplated Mark's story for a few minutes and then touched the com-link, saying, "Sara, I would like to request assistance to renew the contract."

"Very well," Sara said, "I will set up a regular one-year renewal."

"No," Chantal said quietly, pausing for a moment. Make it three years. Thank you, and goodnight, Sara."

Chantal whispered to herself, "You have helped me, Captain."

Mark, entering the bridge, said, "Hello, Sara, I have another coordination instruction for you after our next stop."

"You might like to know that Chan renewed her contract for another three years. Do you approve?" Sara asked.

"Three years?" Mark said, smiling, "I am glad to hear that - approval granted!"

Chapter 6

"Ah, man, I am soaked!" Bill exclaimed as he unloaded dripping containers from *Marvel*. Won't this monsoon ever let up?"

Mark zipped his rain gear up another notch and tried to stay dry under the eaves of the ship. "Not for another month at least. Enjoy your three-day layover, Bill," he added, chuckling.

"What are you up to for the next three days?" Bill asked Mark, who was straining with yet another heavy load of supplies.

Mark replied discreetly, "I am going to kill two birds with one stone," Mark smiled. "I'll be back in three days!"

Leaving the docking area, Mark raised the hood of his rain jacket over his head as he stepped out into the driving rain. *Walking in a warm monsoon means I won't be shivering in the rain*, Mark thought as he passed the port gate and waved his hand to the closest skimmer taxi.

"Crystal Park!" Mark requested as he stepped into the taxi.

"In this rain? Never mind, it's not my business," said the taxi driver.

The taxi rose a foot and skimmed along the nearly deserted highway, then exited the main road and traveled down the hillside service road, where it stopped in front of an overhead sign that read, 'Crystal Park'. Mark paid the driver and strode along the gravel path through the open gate, squinting through the driving rain as he found his bearings amongst the myriad of crystal statues and monuments decorating the park. The pathway ended at another walkway surrounding a small, beautiful lake. Taking precedence over the scene in the lake's center was an immense crystal fountainhead cascading with glittering water.

Through the haze created by the gushing water and the heavy rain, Mark found what he was seeking: a bench made entirely of glass facing the fountain. Several large shrubs surrounded the bench, inadvertently creating a semblance of a privacy hedge. The park seemed deserted. Mark sat down on the bench and, glancing at his watch, said to himself,

She should be here any moment.

Sensing another presence, Mark looked up to see a tall, dark individual walking toward him along the pathway. The stranger approached the bench and sat down. In a soft, familiar voice, she said, "It's good to see you again, Mark."

"I am glad to see you too, Mel," Mark responded gently.

Melissa pulled back her hood with one hand, turned to face him, and retrieved his hand.

"I like this spot. It's a perfect place to conduct our business without people taking notice of us," said Melissa.

He replied, "It's too bad about the weather, though. Crystal Park is known throughout this world and many others for this exact pathway in front of us. It has been said that an evening walk with the multi-tiered reflections of light from the fountain heightens the senses and provides a romantic backdrop for any personal or professional meeting." Setting the landscape aside and looking at the water, he continued, "You found the fourth black stone, didn't you?"

"Yes, but it is on another planet, and my people require your services again. There has been another security breach, only this time the Commonwealth authorities have mounted an intense investigation as they look for the black stone, too."

"Do they know what the black stone does?" Mark asked, concerned for his companion.

"No, but they might if they have something that could detect its presence. My people require a third party to provide the groundwork and avoid the Commonwealth as they try to recover the stone. I knew if I asked you, you could assist me in this matter. Who can the Meloson trust to smuggle the stone off-planet and hold it for safekeeping, Mark?" Melissa asked.

"You mean, long enough for me to pick it up for you?" Mark asked. "Where to send it is the easy part. I know Jimmy Burlington, one of my most trusted vendors at the Wayside Station located in this sector. Travelers call it the 'Void'. Most smugglers and merchant traders know where Wayside and the Void are. A third-party smuggler that can be trusted is a tougher one to answer."

Melissa told Mark which solar system the black stone was hidden in.

Mark smiled and said, "I just remembered an 'old goat' - a smuggler your people just might be able to trust."

Melissa, puzzled, asked, "An old goat? What is that?"

Mark chuckled and continued. "I mean that he is a human elder named Lem Broughton with a rebellious and hypercritical attitude. But underneath his crusty exterior, he is responsible and loyal to other travelers and me. And, more importantly, in this case, he responds well to bribes." Smiling, Mark said, "Mel, could you arrange for one of your agents to purchase and ship a small crate of hickory-smoked Pacific Coho salmon for me from Earthside? I think that should keep our smuggler content."

"I think that can be arranged," Melissa answered carefully.

"Tell your agent that our 'parcel' is only an introduction to get our crafty friend's attention. Make sure the gift is labeled, 'From Captain Mark Styler'. After he has checked the 'gift', your agent can negotiate with Lem to have the stone delivered to Wayside, where Jimmy will be waiting to receive it. I can upload the data from my implant if you want it now."

"Go ahead, I am ready," Melissa said as she touched Mark's ring.

"Contact me again when Jimmy has received the stone, and I will make the trip to Wayside to pick it up. I have a business to attend to there." Mark looked into Melissa's face and said, "Once your people get the last black stone, you will be gone too, won't you?"

Melissa placed both hands on his arm. "Since my mate died, you are the only one I have been able to trust. You proved yourself when you risked your life to save mine. I will miss our time together." Changing the subject,

Melissa observed, "You have been to this park before, haven't you? It's gorgeous."

Mark responded, "Kate and I sat on this bench when she was pregnant. That is when we decided to name our daughter Crystal."

"You told me a lot about your family back in Francia's forest. Do you think you and your daughter will get back together again?" Melissa asked.

"I don't know. Crystal still blames me for her mother's death."

"Have faith, Mark. At least you have a daughter. I don't," said Melissa, her voice barely a quiver.

Once more, Melissa pressed Mark's arm tightly, quickly got up and disappeared into the monsoon. Mark watched Melissa's form fade away as the rain and mist enveloped her body, with his jaw dropped open.

Mark sighed and whispered to the darkening sky, "Mel, I am so sorry for your loss."

Shivering and grim as his sodden clothes cooled in the waning light of the afternoon, Mark rose from the bench and quickly retraced his steps to the main gate and the waiting floater cab.

"With or without Crystal, I think that it is time for me to get my life back in order again," Mark murmured as the taxi skimmed its way back to the hotel.

***

The next day, late morning, Mark exited the hotel lobby

as an android in a hotel uniform held the door for him. "Taxi, sir?"

"No thanks. I was just going for a walk. The downpour seems to have let up," Mark said to the doorman.

"I have observed that it usually does near the lunch hour, then returns," replied the android, discreetly pulling the door closed again against the ever-present drizzle.

Mark walked along the sidewalk under the protection of the hotel's canvas canopy overhead. Although he was glad to stroll along without getting soaked, Mark suddenly felt hungry as he sensed a familiar smell emanating from a new cafe along the promenade. A simple one-word sign in the window read 'Pizza'.

"Why not!" Mark thought as he entered the café. As he entered the small storefront, he saw a long, narrow room. A glass countertop with various types of pre-sliced pizza was on the right side of the narrow room. The small tables took up space on the left side of the room.

A petite, black-haired lady in a white shirt and trousers behind the glass counter asked, "What would you like, sir?"

Mark pointed, "I will take several slices of those two and a cup of black coffee." Mark took the tray with his paid order and sat at the last accessible table. The cafe was busy with many other customers taking orders and leaving the premises.

When Mark finished saying grace, he raised his head to see a young girl no older than 17 or 18 holding a similar tray standing before him.

"May I share the table with you?" the girl asked.

"Sure, be my guest," Mark replied.

"It's been a long day coming!" the girl exclaimed.

The code word! Mark realized. "Especially when the day is much too long," Mark responded. "Today is my last layover day, and I was just about to give up on you people," Mark said dryly.

"Sorry about that. I knew who you were the last couple of days, the man with a black stone ring, but I had to shake off a few annoying agents," said the girl.

"I am relieved to meet up with you, though. You must be Berretta!" Mark inquired. "That is an odd name for a lady."

"My mother is a bit eccentric," Berretta grinned. "I think my mother named me after someone named Berretta. My impression from my mother is that he must have rescued her while she was pregnant with me."

"You can tell me if it is none of my business, but does your mother carry out a similar line of work as you?" Mark inquired.

Tentatively, Berretta replied, "It's ok, and sort of. Why do you ask?"

Mark realized that Berretta was likely a new agent and inexperienced, but he liked her as he replied, "I don't think you were named after a person. A Berretta is the name of a handgun that uses projectiles back on Earth."

"Oh, that makes so much sense, Mark. Thank you!" said Berretta thoughtfully.

"You have something for me?" Mark asked.

Berretta reached into her shirt pocket and pulled out a small silver disk, which she discreetly placed beside her coffee.

Mark rotated the black stone on his ring and then drew his hand back. "Thanks," he said.

Beretta's confusion registered on her face. "Don't you want the disc?"

Mark explained his action. "I recorded the data with my ring. You can have the disc back. I deleted the data off your disc, too."

"But won't the Space Port's scanner pick up your ring?"

"No," Mark smiled, finishing up his pizza.

"Interesting, I have a client who would offer a good price for that ring."

"The ring is designed to be person-specific," Mark replied, sarcastically adding, "Sorry."

"A pity," Beretta whispered.

Mark put his cup on the table and said, "It is time for me to leave. I need to check out of the hotel. It was nice chatting with you, Berretta."

"You too, Mark," Berretta replied.

As he left the café, Mark noticed the sidewalk was deserted and realized the rain was back. He lifted his hood over his head, checked his bearings, and turned toward the hotel.

Mark was walking alone when two men suddenly moved from the shadows of a nearby building. They matched Mark's stride, moved beside him, and seized his arms. One of them poked something hard between his ribs. Without hesitation, Mark whirled around to his right, breaking the man's grasp and, with a downward stroke of his left hand, knocked the stunner away.

At the same time, Mark used his open right hand to land a stiff uppercut to the other man's chin. However, the wet, slippery sidewalk caused Mark to lose his balance, falling heavily to the sidewalk.

Mark stared at the other gunman's smirk as the rain dripped down his face. He aimed his stunner at Mark's head while Mark considered his options. Then, all at once, he heard the hiss of a second stunner, and in seconds, both men fell to the sidewalk, insensible. Mark struggled to his feet and realized that Beretta had her stunner pointed right at him!

For a moment, Mark thought it was game over and she would pull the trigger. Then he relaxed as Berretta shoved her stunner back inside her coat pocket.

"Sorry about that," she said. "You seemed to need an extra hand, but now we had better leave this area before they wake up."

"Just a moment, Berretta." Mark bent over the two unconscious men and scanned his Meloson ring over their unconscious bodies.

"What are you doing?" Berretta asked.

"Searching for video recording devices. I don't want their boss to know about me and my connection with you. I found one in his belt clip and deleted it. My ring tells me this other guy has one, but I can't find it."

Berretta knelt beside him. "I have heard that some agents might have a recorder implant attached to their retina."

"You are right. I just deleted this guy's implant with my ring. Are these the thugs you were trying to shake off earlier?" Mark asked.

"Yes, they must have discovered us sitting together in the cafe."

"Why are they after you and now me?" Mark asked.

"I don't know. I had never seen these guys before, but they have been trailing me for the last two days," Berretta replied. "They must have seen me pass you the disc and figured you must have something important."

"It doesn't matter; my ring recorded their identity from their implants. I have a friend who can find out more about them. I would say it's time for you to keep a low profile for a while. Look after yourself," said Mark.

"You, too!" she beamed.

As Mark returned to the hotel, he whistled an old song, 'Raindrops Keep Falling On My Head', as the noise of the downpour drowned out his whistling.

***

"How was your layover, Chan?" Sara asked.

"I am soaked!" Chantal said, in a foul mood. "I have never known a planet where half the world is in a monsoon! I hope the next stop is a lot dryer. Have you heard from Mark?"

"I just received a message from Mark. He will be here shortly and has asked us to prepare for off-world."

"Did Mark have enough of the rain?" Chantal asked with a smirk.

"All Mark said was that he got his two birds," Sara replied.

"Sara, where did Mark get a hold of you? Or is that a secret, too?"

"Sorry, Chan. I don't have Mark's permission to explain my existence. Give Mark a little more time to get to know you better. Someday, Mark or I will inform you how and why this is all kept a secret."

"Mark just boarded the *Marvel*," said Sara. "One moment, Chan. Mark wants my attention in his quarters."

Mark entered his room, sat by his desk and forwarded the ring's information on the two agents he had fought with earlier to Sara while she searched through the 18 billion personnel database.

"I have a hit on one of the two agents," Sara replied as she continued, "Jeff Micro, age 64. A Commonwealth agent since 24 years of age. He was assigned to keep his eye on a possible suspect, Berretta Treves, regarding a Navy security leak."

"What type of security leak?" Mark asked.

"Unknown," Sara replied.

"This is too close a call, Mark. You were almost taken in for interrogation, and *Marvel* boarded. We should leave now just in case they find out about me," Sara suggested.

Mark nodded as he rose from his chair, exited the room, stepped on the bridge, and sat down in his pilot chair.

"Are you prepared to take us off-world, Chan?"

"Drive pile now charged, and Mission Control has permitted us to go off-world."

"Okay, let's go," said Mark.

Chapter 7

Chantal rotated her co-pilot chair toward Mark and asked, "How often do you stop at Wayside Station since it's in the middle of the Void?"

"Once a year," Mark replied.

Chantal knew little about Wayside Station and had to find information about it in *Marvel's* library. Chantal found a text about the Void in the vast library, opened it and read, "The 'Void' lies in sector four near the borderland, an area empty of any stars, planets, or orbs within a radius of 4 to 6 light years. Within the Void lies a huge privately-owned outpost called Wayside Station. Wayside consists of obsolete collections of old space liners, one battleship, and other large and small ships permanently stationed there. Following 2217 and the resulting second wave of migrations from Earth, Wayside Station was constructed, jointly-owned and operated by twelve merchant trade guilds. Legal and illegal activities, trades, and other profitable business arrangements are discreetly carried out there. The Commonwealth authorities ignore Wayside's existence as long as the guilds paid their taxes."

The dispatch office suddenly interrupted. "Wayside control to *Marvel*. You may proceed to dock # 41 as instructed."

"Thanks," Mark replied.

"It's quiet here," Bill said. "Only half of the docks are occupied. I wonder why?"

"We'll find out soon, I guess," said Mark.

"I wish I could come with you," Tim said, disappointed.

"Sorry, Tim. I did promise your father I would not let you on Wayside," said Mark.

"Yes, I know. Dad made me promise, too," Tim said in a subdued, low voice.

"Whatever you do after your apprenticeship is finished will be your business, Tim. I am sure your father assumes that by then, you will be mature enough to recognize any shady characters," said Mark.

"That's only a year away!" Tim smiled.

"You think your girlfriend Theresa will be waiting for you on Wayside, Bill?" Chantal asked.

"I hope so. We should be planning for our retirement together very soon."

"Are you going to go through with your retirement plans?" Chantal asked.

"Chan, I am now 100 years old. I have only 20 years left before this body hits the threshold and breaks down. Theresa and I are the same age, we get along, and we have decided to enjoy our remaining years together while our health is good." Bill paused, took another look at the computer screen, and added, "But something doesn't look right here."

"Do you think it is too risky to board Wayside Station?" Chantal asked.

Bill replied, "Don't get into trouble with the locals and be cautious when approaching any vendors that give you a friendly smile. I think you'll be okay. I suggest you stick close to Mark until you get the hang of this place. There are many interesting things to see but plenty of opportunities for trouble, too!"

"What is your business here, Mark?" Chantal asked.

Surprised that an android would ask such a question, Mark replied, "I have an appointment with an old acquaintance."

"An old friend?" Chantal inquired tactfully.

Mark smiled and added, "Actually, I wouldn't say that. Let's say that we have a mutual respect!"

"*Marvel* to Wayside Control, docking in ten seconds," Chantal announced from the co-pilot chair.

As the countdown continued, Chantal watched as the docking tube extended and locked onto the *Marvel*. "Docking is secured; welcome aboard, Captain Styler. Please follow the security scan instructions near the boarding entrance," announced the dispatcher.

"Copied and out," Mark replied. "Sara!"

"Yes, Mark?"

"After we are off-ship, lock the *Marvel* down tight."

Mark continued, "Mason, I want the drive piles kept fully charged just in case we need to escape from here quickly."

"Yes, Captain," Mason replied from the engine room.

"Are you expecting trouble, Mark?" Bill asked.

"I'm not sure, Bill, but I agree that there is something not right about this place. It's too quiet around here. It's better to be safe than sorry. I don't want any mistakes like the last time."

Bill nodded his head in agreement.

"What mistake was that?" Chantal asked.

Mark didn't answer as he rose from his chair. *Another secret*, Chantal thought.

"Listen up, everyone. Unless you hear from me, report back to this ship by 05:00 hours for reassessment," Mark announced.

"I am off!" cried Bill as he rushed to the exit.

Mason checked to ensure the drive pile was fully charged, then quickly left the engine room and followed Bill to the ship's exit.

Chantal rose from her chair and quickly stepped in line beside Mark as he left the bridge. "Where are we going on Wayside?" she asked curiously.

Mark stopped in his tracks and said, thoughtfully, "You sure ask a lot of questions for an android."

"Never mind. Let's go."

Alarmed, Chantal thought to herself, *Keep it cool, Chan. Otherwise, Mark will get suspicious!*

As Mark and Chantal entered the Wayside Station entrance hallway, a voice from the station computer announced, "Stand by for a security scan!"

"Ready," Mark said as a transparent blue shield hovered over Mark and Chantal. Moments later, the scan completed, the station's computer announced, "No weapons or bio subjects are registered. You may proceed."

"Chan, look all you want, but stick close to me." Mark walked down the several levels of passages before leaving and entered several more joined ships.

As they walked through the crowded room, Chantal looked about astounded. She had never seen so many different aliens and humanoid species in one place. Regardless of what Mark and Bill had said earlier about safety, everyone showed respectable consideration for each other. Chantal then realized why. She noticed armed security personnel with Wayside's emblem on their uniforms mingling amongst the locals, with the apparent role of keeping peace. As Mark entered a silver-gray metallic interior, Chantal realized this was an old battleship.

A hovering red ball in front of Mark said, "Mr. Styler. Mr. Rand has changed his office. Please follow me as I direct you to his new office."

Chantal tagged behind Mark as he followed the red ball, which moved silently along the gallery, traveling several levels up and along multiple passages, entering an elaborate reception lounge. Inside were three security guards dressed in surprisingly modern business suits and ties. The closest man stood in front of Mark with a handheld scanner, the other behind the desk, and the third in front of the carved wooden door.

"Captain Mark Styler? Please hold your arms up above your head," the man said as he scanned his body. "You're clean," he said. "You!" said the security man in a firm voice as he pointed his finger at Chantal. Sit over there!" Mark nodded his head to Chantal as she walked to the left side of the room and sat on the chair.

"Mr. Rand is expecting you," said the man behind the desk. The third man opened the door as Mark entered the office.

In the office of a collector, everything in the room was full of century-old cherry wooden furniture with paintings hanging on the wall. There was little doubt that they were original paintings and not reproductions. Numerous artifacts, some unfamiliar to Mark, could be seen on shelves or platforms. Mark approached the most enormous walnut desk he had ever seen. Standing behind the desk was a short, slim man with white hair. Mark knew that Stacy had been born with white hair. Stacy was beyond the threshold age of 120 years, kept alive by the latest medical lab-constructed organs and drugs. Behind Stacy on the wall was a large painting of a medieval Germanic man, one of Stacy's ancestors. Mark had known Stacy for nearly 40 years as a merchant trader and a friend or, at the very least, a shrewd business associate.

Stacy rose from his chair behind the desk and held his hand out toward Mark.

"Hello, Mark," the man said as he held his hand. "Good to see you again."

"Thanks, Stacy," Mark said as he shook his hand in return.

Mark sat down on a leather chair in front of Stacy's desk.

"Is all well with you, Mark?"

"The past year has been good," Mark replied.

"Glad to hear that, Mark! How is Bill?" Stacy asked.

"Looking forward to being with Theresa again."

"Do you think they will ever retire and settle down?" Stacy asked with a rueful grin.

"I think so, perhaps in a few short years," Mark responded.

"Glad to hear it," Stacy replied, "Theresa has been almost like a daughter. Bill is a good man for her." With a quick shrug, Stacy moved on. "Let's get down to business, Mark. This room is secured. Do you have something for me?"

Mark rotated Melissa's ring and responded, "Yes, I do. Ready to transmit data." He downloaded the data to the memory bank embedded within the desk.

"It's amazing that my best scanner can't detect your ring. Could you enlighten me as to where you attained this ring?" Stacy asked.

"No, I am sorry, Stacy. I made a promise not to disclose that," Mark replied.

Stacy shrugged his shoulders and said, "Fair enough, Mark. Now, let's see what you have got here." With that, Stacy said, "Open monitor." A hologram of a monitor appeared above the desk. Stacy muttered some words as the data and images materialized on the monitor. Moments later, Stacy sighed and looked troubled.

Mark was shocked to see how quickly Stacy seemed to age before his eyes as he leaned back slowly in his chair and closed his eyes.

"I assume it's not good news?" Mark asked.

"No, not good news at all," Stacy whispered quietly.

"What is it?" Mark asked.

"The data you forwarded to me is at the highest level of the Commonwealth Security Office. The good days of doing business will be over soon—or at least for us here. The Thracian Empire is at it again, or they soon will be."

A shiver ran up Mark's spine as he desperately tried to block the five-year-old memories that came flooding back. Mark took a deep breath and said too forcefully, "Why? I thought we had a treaty?"

"We do, but the Thracian Empire today is still a classic dictatorship of Klan warlords. As the name suggests, klans are political family units of collective, cooperative warlords, whereas clans are smaller, family-run units, united by actual or perceived kinship in organized crime. When humanity's nearest settlement encountered the Thracians, they were annihilated before the Commonwealth was able to stop them. Their Thracian society is over 2000 years old. Consider that in light of the

300 years our culture can celebrate. The Thracians' technological advancement was often hindered by the warlord Klan's bloated bureaucratic honor system used to stabilize their society. Selective new technologies must be copied and shared amongst the rival klans to prevent unfair advancements between rivals. The rest are kept hidden in their vaults controlled by the elder warlords. In this way, both the political and honor systems are tightly maintained within Thracian society. In comparison, Commonwealth society and its technology are not as rigidly controlled or suppressed. That's why the Thracian Empire's expansion has been limited to only 20 light years within 150 star systems. However, the Thracian world has only a 12 G-class star system, much like Earth's Sun, Sol, for example, that is suitable for life support. Conversely, the Commonwealth is only three hundred years old, spread out across four sectors, 50 light years, and consists of 1875 stars and 65 G-Class Suns suitable for life support." Stacy paused for a few seconds, then continued, "However, during the Thracians' last 2000 years of conflict, they have bred warriors that are bigger, faster, and smarter than any of the Commonwealth Navy combat personnel. We won the war 50 years ago because our warships were faster and better equipped."

"Something has changed," said Mark.

"Yes. When the Commonwealth defeated the Thracians and threatened to destroy their home world, a treaty was established in which the more elderly warlords were to be retired and replaced by younger, less experienced, and aggressive generations. A further peace accord was established in which a buffer zone of 10 light years was created between the Thracian Empire and the Commonwealth."

"No military, civilian ships or settlements from either side are to exist inside the Buffer Zone. Many travelers also call it the Borderline or Borderlands. A joint commissionaire committee was established for both sides to settle any disputes in the two outpost stations. Firearms are limited to sidearms."

"Except for the G-Gun," Mark said bitterly as he tried to suppress his painful memory.

Stacy paused for a second, nodded, and said, "Yes, except for the G-Gun."

"Go on," Mark said irritably.

"It has never gone well with the older warlords being removed and replaced by the younger ones. During the last 50 years, some members have gradually improved their status to become personal advisors to the younger warlords, leading to a reform movement. There has been a renewal of the old ways of the colonial empire among the newer warlords, somehow, they must have found one of the Commonwealth's destroyed battleships and copied our superior warp drive and weapon systems. Before the Commonwealth left the region, spycams were installed in the Thracian region, which has shown an increased level of construction of new battleships on two planets. I don't need to tell you what is on their minds next."

"Yes, I know their intentions are aggressive expansionist movements," said Mark, adding, "Should the Thracian prepare to invade the Commonwealth's four sectors, has the Commonwealth been making any preparations to counteract this?”

“A Commonwealth pre-invasion!” said Stacy.

“When?” Mark said, shocked.

“That part, I don’t know yet. But I do know this. Wayside is too close to the Borderline, and it is low on the Commonwealth’s protection list.”

“And what do you hear among the traders, Stacy?”

“After filtering out the rumors circulating among the rumormongers, I would say we don’t have much time left. The tensions in Sector Four are increasing. Some reports have been of an unknown scout type, and other ships are cruising along the borderline. And if these rumors are true, the Commonwealth is blinding to the scout violations.”

“Why is that?” Mark asked.

“To keep our neighbors, the Thracians, off-guard? I don’t know yet. You must have noticed that the traffic around here is down considerably.”

“I have indeed. So, what are your plans?” Mark asked.

Stacy replied, “Leave the Void, of course. Travel far from this sector then lay low for a while.”

“What about the rest of the trade guilds?” Mark asked.

“Most of them are at least half my age, and I am sure they will be young enough to regroup and start over somewhere else. As for me, even though I am 15 years beyond the threshold, the doctors on Wayside think they could keep me alive another 10 to 20 years." Stacy added ruefully, "Unless the Commonwealth medical staff have a major medical breakthrough and solve the mystery of the threshold."

"Maybe," Mark smiled in return.

Stacy made an effort to focus his eyes on Mark. "You believe your God has set the age limit at 120 years, don't you?"

"There are good reasons God does anything we don't understand, Stacy."

"You might be right, considering that current knowledge about the brain's capability for memory capacity suggests that life should be attainable upwards of 800 years or more. Maybe that's why those in the book of Genesis live up to 900 years. 900 years. Now, wouldn't that be interesting? Oh well, maybe in the next generation from now," said Stacy.

"What about you, Mark, now that you know about the Thracians?" Stacy asked.

"I still have several obligations ahead of me. I have two deliveries, and after that, I don't know. What would you recommend?" Mark asked.

Stacy considered momentarily and made a few adjustments to the hologram monitor. An image of the four-sector star system appeared. He explained, "This yellow dot is Sector Four, where we are now, near the Borderline. The blue-dotted outline in this direction on the map is the Commonwealth's four sectors. The other blue dot outlined below Sector Four is the new frontier, the Fifth Sector, about 45 light years away. The merchant guilds are moving to the Fifth Sector to avoid bloodshed and to renew trading partnerships again."

Mark engaged his implant to record the data shown on the hologram. "Thanks, Stacy!"

"You staying long, Mark?" Stacy asked.

"Just long enough to finish my business here," said Mark.

"Alright, I will see you out," Stacy said, rising.

Mark didn't move from his chair. Folding his hands together, he waited.

"Mark, I was going to do it," Stacy said innocently.

"I was sure you would." Mark smiled sardonically.

Stacy sighed, leaned back in his chair and touched a small panel on his desk. Swiftly, he spoke into the embedded microphone, "Jones!"

"Yes, Mr. Rand?" said a familiar voice which Mark remembered belonged to the man behind the desk in the lobby.

"Transfer 700 credits to Captain Mark Styler's account," Stacy requested.

"Yes, Mr. Rand."

"No, it's 1700, Stacy!" Mark cut in.

"Ah, yes. Jones, make it 1700 credits", Stacy said.

"Completed, sir."

Glancing at his ring, Mark turned the black stone clockwise and said, "Ship computer!"

"Mark, nothing can transmit out of this room," said Stacy.

"Yes, Captain," Sara responded.

Stacy gasped at the ring.

"Did I just receive 1700 credits in my account?"

"No credit transfer received yet, Captain Styler," Sara responded.

Mark turned toward Stacy and firmly said, "Stacy, no more games, please!"

Stacy acquiesced and, in a low voice, responded with, "Mr. Jones, please forward 1700 credits to Captain Styler. Now!"

"Captain Styler. Your account just received 1700 credits," said Sara.

"Thank you - out!" Mark turned the black stone back to the original setting.

"That is one amazing ring you have got there, Mark! You don't want to sell it, do you?" Stacy inquired.

"I am very sure," Mark replied with a grimace.

"Oh, Mark, one more thing," said Stacy.

"What is it?" Mark asked.

"I have always enjoyed doing business with you, so here is a tip: stay away from the Borderline; a few small ships have been reported missing lately in that vicinity."

"Thanks again," Mark said, rising and shaking Stacy's hand.

Chantal quickly rose from her chair when Mark entered the lobby. "Let's go!" he said as Chantal rushed to catch up with Mark and asked, "What's next, Captain?"

"I'm going to give you a personal tour of the trader's open market. Afterward, I'm going to meet up with another friend."

As Mark and Chantal walked toward several more vessels, they entered an exceedingly long tunnel that led to another ship. Chantal stopped in her tracks and looked beyond in amazement. The vessel was a vast 500-foot-wide sphere. Chantal followed Mark as he strode across the platform, passing by many booths. Many were makeshift booths with simple countertops or shelves. The others were more professionally designed. Several vendors moved as close as possible toward the platform, pacing Mark and Chantal and calling out to the pair as they advertised their goods. Chantal looked up at the sudden laughter and shouting and realized that the market comprised at least six floors. There was a 30-foot-wide walkway that circled the interior wall and contained a platform-style elevator conveying tradespeople between the causeway and each floor. Chantal had seen trading markets before, but this complex was huge.

"What was this vessel originally used for?" Chantal asked Mark as they made their way through the main floor of the agora.

"It was a refuelling station for the first fleet of starships from Earth!"

"Even though this is much bigger than anything I have seen before, there are a lot of empty booths," Chantal remarked.

"I will tell you why as soon as we return to the *Marvel*. Follow me." Mark led Chantal to the elevator.

Chantal gazed upon the scene below as the elevator rose to the fourth floor. Quickly, they rose and just as quickly exited the elevator. Mark and Chantal walked toward the circular causeway and into the crowd of marketgoers as the doors closed with a whooshing sound behind them.

Mark approached a booth with a sign that read "Burlington." A heavy, dark-skinned man sat behind a desk before a small tent gazebo. As he looked up at Mark and Chantal, without saying a word, he rose and silently moved behind a curtain hanging from the gazebo. Mark sat in a small chair in front of the desk and waited. The man returned, holding a small plain box and placed it on the desk. As he was about to open the box, the man looked up and saw Chantal. He froze.

"Not to worry, Jimmy. She is one of my employees, and I trust her," Mark reassured him.

Astonished, Chantal thought this must be the first time she had heard anyone say that about her.

Jimmy lifted the lid of the box to expose its contents. "Is this what your client wanted?" he asked nervously.

Chantal gasped. A shining dark stone that was small enough to fit in the palm of her hand was nestled into the heavy folds of silk lining the box. It was a much larger version of the one on Mark's ring. In wonderment, Chantal watched as Mark touched his ring to the much larger stone.

Mark was jolted as a warm sensation shot up his arm, then subsided as he pulled it away before he moved the ring away from the contents of the box.

"That will be two and a half," Jimmy said with a sideways glance at Chantal.

"My client is only prepared to pay two," Mark replied.

Jimmy sighed, "Things are getting quiet lately. Alright, two! It's getting harder to do business. I am thinking about shutting down my shop," Jimmy replied petulantly.

"Where would you go, Jimmy?" Mark quietly asked as he transferred the two thousand credits over to Jimmy's account.

"Back home, I am getting tired of this line of business. I think I will lay low and retire after a while."

"Good luck, Jimmy," Mark proffered as he removed the stone from its concealment and dropped it into his waist pouch. Mark checked to ensure the pocket clip was secure then sealed it. He gestured to Chantal and rose from his chair.

"Wait, Mark! I have got a business proposition for you." Jimmy lifted his arm to stop Mark from leaving the stall.

"What kind of proposition?" Mark asked.

"A relocate," Jimmy replied.

"Not interested," Mark said as he stood up.

"It's legal!"

"Still not interested," Mark replied.

"Two and a half up front and two and a half upon delivery."

5000 credits! Mark thought.

"Where to and why me?" Mark asked.

"I can't say unless you're committed."

"Goodbye, Jim! Let's go, Chan."

"Alright, but you're putting my head on the line, Mark."

"I won't tell anyone, Jimmy, you know that!"

"I know, that is why I am offering this sweetheart of a deal only to you. Sit down, Mark, and I will tell you what I know."

Jimmy leaned forward, folded his hands on the desk and, looking straight into Mark's eyes, said, "There has been a lot of trouble between two family clans of Italian descent and their rivals from Planet Romulus in the Sector Four solar system of Stella. To promote unity among the family clans, there has always been a binding marriage arrangement between the Russo and the Greco families.

"The Commonwealth won't like that," Mark said. Marriage arrangements tended to create hostility and jealousy among the elite clans in the sectors and were banned by previous governors of the ruling sectors.

Jimmy curtly responded, "What the Commonwealth doesn't know won't hurt them, Mark. After a few rejected offers, the families decided on a distant young cousin arrangement from an influential school on the Planet Remus, a nearby sister star, Stelletta, in Sector Four, at just a few light years' distance.

"Remus," Mark smirked.

"You know it?" Jimmy asked.

"I had some personal business with Remus before, and they have a reputation for supporting a very arrogant population, and I have no admiration for them. Go on," said Mark.

"Everything needs to be kept a low profile because of the Commonwealth authorities and the other clan's rivals on Romulus."

"Why would the other family clans be a concern?" Mark asked.

"Power and influence and a greater share of profits," Jimmy replied.

"Some things never change; money is always involved somehow," Mark muttered.

"What's that?" Jimmy asked.

"Nothing. Keep talking," said Mark.

"A yacht left Remus for Romulus with this cousin as a passenger, but there was an accident. A malfunction in the warp drive caused the ship to wander off course. When it dropped out of warp space, all the crew had left was the basic pulse drive. Somehow, the ship made it safely to Wayside."

"How far is Romulus from here?" Mark asked hesitantly.

"About 12 light years," Jimmy replied.

Mark crossed his arms. "Here's how it is, five up front; he obeys all the ship's rules and follows my orders. And he comes alone."

"I am not sure, Mark, that is a very unusual demand," Jimmy responded.

"Jimmy, I have been ripped off by a Remus client before and had a tangle with a Remus bodyguard, which left me with one bad memory. That's a mistake I won't make again. If he wants a ride, he must be at dock 41 at 08:00 hours tomorrow, Wayside local time. We leave by 09:00 hours."

"I can't promise anything, but I will see what I can do," Jimmy said doubtfully.

"Okay, see you tomorrow, Jimmy, at the boarding dock," Mark said as he rose from his chair.

Mark nodded to Chantal and said, "Now I am finished, Chantal. I hope you have enjoyed our little excursion. Now, let's get back to the ship."

***

The following day, Bill and Chantal were ready and waited at the boarding dock at the allotted time. The minutes ticked by. Just as Bill became annoyed, what he thought would be a 'no show' arrived.

"Captain, our guest has arrived and has brought others with him," Chantal announced over the com-link.

"I'll be right down!" Mark said. Moments later, he and Misty arrived at the boarding dock.

"Captain Mark Styler?" asked a captain in Remus' military uniform who stood before Mark.

"That I am. How may I assist you?" Mark asked sarcastically.

The captain, slightly put off by Mark's remark, retorted shortly. "I am Captain Moretti of the Remus Imperial Navy, and I am to assist Marion De Luca to Romulus."

Mark turned his head to face a tall but somewhat heavy-set young brunette. "I thought we were escorting a man?"

"A deliberate act on our part is for security reasons." The captain replied. "May we come aboard?"

"No!" Mark said.

Captain Moretti turned to Jimmy. "But we have an agreement!" he sputtered. Jimmy visibly quailed in the face of his authority.

"It is okay, Jimmy. I can take over from here," Mark said, continuing, "Our agreement is for one person only and full payment up front!"

The Remus captain's face turned cherry red, and he exclaimed, "This is rubbish! I will not permit this, Captain Styler!"

Marion stepped in front of Captain Moretti, facing him. "We did agree to his terms, did we not?" The Remus captain murmured his assent.

Marion turned to face Mark. "Captain Styler, we agree to your terms. May one of your crew members show me my room?"

"Chantal!" Mark called.

"Yes, Captain Mark. Marion De Luca, please follow me," requested Chantal.

"Thank you," Marion replied. As Chantal led her away from the delegation, Marion suddenly stopped, turned to Mark and asked, "You have a canine onboard?"

"No, Miss De Luca. We have Misty," Mark replied.

"Oh," Marion said with a confused glance, "I see!" She turned back and continued to follow Chantal into the tubing dock.

"Captain Styler, if anything should happen to Marion De Luca," Captain Moretti threatened.

"Enough, Captain! It is done! Marion de Luca is safe here. I require payment," Mark demanded.

The Remus captain seemed ready to explode, then thought better of his actions and turned to his waiting lieutenant instead. "Pay him!" the captain cried as he turned and stomped back to the passageway leading to Wayside.

"Transaction completed, Captain," Sara announced.

As the Remus delegation moved to follow their captain back to Wayside, Mark realized that two pieces of luggage had been left behind. "Ah... Bill?"

"I know," Bill grumbled as he picked up the bags and carried them through the tubing dock and into the *Marvel's* passageway.

Misty looked up at Mark and panted. She whined plaintively and shook, sending fur flying. Mark laughed and said, "I know, Misty. People are a strange breed."

"This is your room, Marion De Luca. I think you will find everything here to your satisfaction," said Chantal.

"Thank you. I am relieved. This room is better than I had expected. And where are my belongings?" Marion asked.

"Right here, Miss De Luca!" Bill announced.

"Thank you. Please put them over there. You require a tip, I suppose?"

Bill rolled his eyes, "That won't be necessary," he answered abruptly, then left the room. "A tip," Bill muttered, shaking his head.

"Ship protocol requires that all passengers and crew members be on the bridge before departure, approximately 15 minutes from now," said Chantal.

"That sounds like fun; I'll be ready by then," replied Marion excitedly.

"I strongly recommend that you be on the bridge no later than 10 minutes from now. There are consequences for lateness. The captain will be quite angry," Chantal advised.

"Oh dear, we mustn't have that, must we?" Marion responded.

Chantal dipped her head respectfully, left Marion's quarters and started for the bridge. Chantal thought, *This trip would be just like my old cruise liner days.*

Once upon the bridge, Chantal took her place in the co-pilot seat. "What do you think of our guest?" Mark asked, turning toward her in his chair.

Chantal said, "Miss De Luca seems quite egotistical and reminds me of our VIPs on the liner in my previous employment."

"Remus citizens are well known for that, Chan. The royal families have a reputation for the same arrogant characteristics," Mark replied.

"What do you know about the royal families?" Chantal asked.

"I married into one," Mark muttered under his breath.

Chantal looked at Mark, not sure if he was serious.

Sara announced, "Separation begins in five minutes." Just then, Marion nonchalantly entered the bridge area.

"Oh, there you are, Captain Styler. I must congratulate you on your fine ship. Where is the passenger seating accommodation, please?"

"Just select any of the seats behind me," said Mark.

"Have they been treated with anti-bacterial wash?" Marion asked.

"I can assure you, Miss De Luca, that Bill has kept everything spotless in response to your requests and satisfaction," Mark replied. "While we are preparing for separation, I would appreciate it if I may have your cooperation in silence."

"Oh, you may. Please continue, Captain!" said Marion.

Mark sighed in distaste and spoke over the com-link, "Mason, are we set to go?"

"Everything is online, the drive pile is fully charged, and we are prepared to go," Mason responded promptly.

Chantal engaged the repulsion as the *Marvel* separated from the docking tube. Mark took over flight control, engaged the impulse drive, and *Marvel* quickly left Wayside behind.

As there weren't any large mass objects anywhere that might interfere with the jump drive, Mark could call Sara and say, "Open the warp field." A circular rainbow appeared before the *Marvel.*

Sara announced, "Romulus homing sensor detected and locked on! Romulus is 12.4 light-years away, and our time is estimated at four days and two hours."

Mark engaged the warp drive, and *Marvel's* image stretched in a long streak of light as it entered warp space and disappeared.

Chapter 8

"Captain Styler, while we are in warp space, may we have dinner together?" Marion added, "Bill, please prepare two meals for Captain Styler and me."

Mark heaved a sigh and said, "Go ahead, Bill. Please put together two meals."

"On my way, Captain," Bill smirked. "Two meals coming up!"

"I think this is going to be a long four days," Mark quietly muttered. "Chan, please give our guest a tour of our ship while I check everything over."

"Yes, Captain." Chantal was puzzled by Mark's request. *Why does he want to check things? Sara usually does that operation,* she thought.

"Miss Marion de Luca, if you would follow me," Chantal said as she led her away from the bridge. Misty was determined to tag along with them.

When Chantal and the guest were off the bridge, Mark said, "Sara, I think this will be a long four days. Marion has a typically annoying Remus personality. I think there is more to Marion than she is letting on, so I told Misty to keep an eye on her."

"Do you want me to give you Marion's profile?" Sara asked.

"You have her profile?" Mark said, pleasantly surprised.

"As soon as I realized who our guest was, I hacked, or I should say, I contacted Remus' private ship library."

"Bad girl, Sara!" Mark said, grinning.

"Also, I found some interesting background data on our guest. From what I have collected so far it's quite interesting. There is a power struggle between the two main family clans, Russo and Greco, on Planet Romulus, both Italian descendants from Earth," said Sara.

"That I already know. Go on," Mark said.

"One side of the clans wants to break away from the Commonwealth, and the other wants to stay," said Sara.

"The old good guy and bad guy scenario," Mark said.

"Yes, but who is the good and bad?" Sara asked.

"Well, I would think those who want to stay with the Commonwealth are the good guys!" said Mark.

"Most people would have said the same thing, but not this time," Sara countered.

Mark sat straight in his chair, "Why not this time?"

"The Russo are known as reformists that want to break away from the Commonwealth, and the Greco want to form political associations with the Commonwealth. The Russo demanded their right to political freedom and maintained the android ratio among the Romulus population. At the same time, the Greco wanted to increase android slavery and rule the other political parties with an iron fist, which the Commonwealth would allow."

"Why would the Commonwealth do that?" Mark speculated.

"The Romulus found a minor planet with the highest concentration of rare metal ever found neatly under their jurisdiction. The nearly unlimited rare metal production for the warp drive would allow a much lower operation cost for the Commonwealth and everyone else."

"What does this have to do with the androids?" Mark asked.

"The minor planet has no atmosphere and has an average temperature of minus 250 degrees. The workforce is not available for mining production, but the Romulus could increase the construction of the androids with custom knowledge programs to do the mining and construction. With the Commonwealth's security needs, it would benefit both the family clans and the Commonwealth."

"So it is all about wealth and supremacy again," an idea that left a sour taste in Mark's mouth.

"Yes, it is, Mark, but the information I have gathered gets more complex."

"Why is it complex?" Mark asked.

"Both clans want to compromise, solidify the families' union, and share power and wealth with the Commonwealth."

"Let me guess, an exchange of marriages as an offering of goodwill to each clan and their sub-clans. That means marrying cousins in addition to sons and daughters," said Mark.

"Correct," said Sara, "But one major subgroup is holding out."

"Is it the family clans connected to none other than our young Marion?"

"Right again, Mark," said Sara.

"What is so special about Marion?" Mark asked.

"That's just it, Mark. Very little, as Marion is a mystery. We have only four days to figure this out. One more thing, Mark. You are not going to like this one. I discovered that the Commonwealth authority has tapped into Wayside's security system and cams. We should keep a tight security detail on Marion. Let's leave the crew out of the picture, but Chan might be useful."

"Can we trust Chan on this one?" Mark asked.

"Mark, Chan is more remarkable than you may think, and I trust her."

"Why?" Mark inquired.

"Company coming," said Sara.

Mark rose from his chair as Marion and Chantal arrived on the bridge.

"Captain Styler, I am impressed with your little ship, and your chef just informed me our meal is ready to be served! I would be delighted to learn of your past adventures!" Marion wrapped her arm around Mark's.

"Certainly, Miss De Luca."

Marion led Mark to the ship's mess deck. Misty followed close behind as they left the bridge.

Tim meandered near to Chantal, smiling. "It seems Marion has snagged the captain."

Chantal wasn't pleased with how Marion had touched the captain's arm. When she realized her emotion, she was puzzled by her reaction. *What is going on here? The captain means nothing to me*, Chantal thought. *This isn't reassuring. Maybe I should have my vital systems checked out. I am long overdue for one. I will do a self-diagnostic check in the sick bay when I am off shift.*

Chantal walked to the mess deck to pick up a cup of coffee and overheard Mark and Marion's conversation.

"Captain Styler," Marion laughed, "That is such a delightful and amusing story."

Chantal watched how Marion stroked the back of Mark's hand. *Marion is seducing my captain! Mark seems to be enjoying his meal with Marion.* Chantal thought, *I don't see what's so funny about his story.*

Mark was feeling strange and lightheaded. He looked at Marion's brown eyes. *How long has it been since Kate's been gone?* Mark wondered—*five long years. I must be lonelier than I thought.*

Marion asked, "Captain Styler..."

Mark cut in and said, "You may call me Mark!"

"Mark, I like that name. Mark, we have such a short time to get to know each other better," said Marion.

"Sounds like a great idea, Marion, Miss De Luca," said Mark, struggling to clear his mind.

Misty gave a long whine from her throat. Marion reached over the table to pick up Misty, but she snarled in return.

"Marion, excuse me, but I think three's a crowd." Mark took Misty into his arms and carried her to Chantal.

"Chan, I don't know what has come over Misty. Take her back to the bridge for me."

Misty was still growling when Chantal carried her back to the bridge.

"What's wrong with Misty?" Sara asked.

"I don't know... Misty seems to act this way toward Marion."

"This is out of character for Misty," said Sara. "Chan, when you are not on shift, I recommend arranging to have Misty checked over in the sick bay. When Mark is free from Marion, he must talk to you about her."

***

Several hours later, Chantal leaned back in her chair. "I need to stretch my legs. Sara, I'll be back in a few minutes after I get another cup of coffee."

Chantal walked through the gallery and bumped into Marion as she stepped out of her quarters. Marion grabbed Chantal's arm, but her touch repulsed her and she briskly removed her arm away.

"I am so sorry, Chantal."

150

Chantal nodded her head and turned to walk away from Marion.

Back on the bridge, Chantal sat in the back row and closed her eyes to calm herself down. *What's wrong with me?* Chantal thought. "Sara, have you found anything strange lately?"

"Stand by." Moments later, Sara replied, "Nothing unusual since we left Wayside. Is there something wrong?"

"I am not sure, Sara." Chantal was puzzled by her peculiar nauseating sensations.

Mark boarded the bridge with a giant grin on his face. Misty started whining again.

"Captain, permission to leave the bridge!" Chantal requested.

"Permission granted," Mark said as Chantal quickly left the bridge.

What's with Chan? Mark thought to himself. *Oh well, it's time to start my shift!*

Chantal entered the fitness room and stepped on the treadmill. *What's with that captain of mine and that silly grin on his face? 'Oh, I am so sorry.' Marion said that to me. I couldn't stand it when she touched me!* Chantal banged her fist on the stop button. *Touch me!* Chantal stepped off the treadmill and ran down to the sick bay. Misty stepped aside as she ran for the sick bay. Misty followed closely behind Chantal and then hid behind the lab counter.

Chantal lay on the analyzer bed and commanded, "Sick Bed, Computer!" Chantal shouted, "Perform a full-spectrum scan!"

After the computer had scanned her and her blood samples were tested, the computer announced, "Scan completed. All vital organs are functioning normally, and an unusually isolated brain pattern is detected. Blood tests are still analyzing."

"What unusual brain pattern?" Chantal asked.

"The archipallium part of the brain stem," the computer replied.

Chantal remembered that this was the location of the brain's mechanisms for aggression and repetitive behavior.

The computer said, "Blood test analysis completed. There is an unknown subject compound detected."

An unknown compound, Chantal thought. "Copy the subject compound and perform a simulated action for known brain functions," said Chantal.

"The required instructions need assistance from the main lab computer to perform the operations," said the computer.

Sara could help, Chantal touched her com-link and said, "Sara!"

Sara replied, "Yes, Chan."

"I am going to download my medical data to you. I want you to perform a simulated action on the data as instructed," said Chantal.

"Data received. We are analyzing the medical data. The analysis is complete. This is very interesting, Chan!" Sara said.

What! You have finished already! Who or what are you? Chantal contemplated. "What did you find, Sara?" Chantal asked.

Sara replied, "There are two separate designer chemical compounds. The first compound creates a daydream effect, subjecting the patient to receive an additional chemical compound in a different area within the brain.

"What does the next chemical compound do?" Chantal asked.

"It creates a hypnotic suggestion. The possibilities are unlimited," Sara replied.

"That may explain why I experienced a strong repulsive reaction to that drug. I think Marion's hand somehow injected some of the drugs by touching Mark's skin. But why did this drug negatively affect me?"

"My guess would be because of your modified brain structure," Sara replied.

"We need an antidote for Mark," said Chantal.

"Done!" A syringe slid out from the computer desk console.

"You are amazing, Sara!" said Chantal gladly.

"Thanks!"

Chantal placed the vial in her sleeve pocket.

"It will take about a minute to neutralize the drug's effects on Mark. Even so, he will probably be confused about what has happened to him," said Sara.

The room lights were replaced by emergency luminosity. The auto P.A. voice said, "Ship computer is offline. Emergency backup is now in force! Please stand by for further orders."

Just then, the door opened, and Mark and Marion stepped in. "Mark, seize your android and immobilize her on the sick bed." Mark was indifferent to Chantal's protest as he restrained her and placed her on the sick bed.

Mark commanded the computer to immobilize Chantal's muscles, which paralyzed her.

Bill was concerned when he heard the commotion down the gallery. He rushed to the sick bay and entered the room, but Marion grabbed him and hurled him to the opposite wall. Bill's head hit the wall, and he fell unconscious to the floor.

Tim and Mason arrived seconds later as Marion took hold of Mason's throat with one hand and squeezed it until he blacked out and fell to the floor.

Mark took hold of Tim and pinned his arm behind his back lifted his body onto the other sick bed and commanded the computer to immobilize him.

"When you finish with the boy, tie up Mason and then stand by. I'll be back in a few minutes," said Marion.

Mark searched the room, found the elastic bandage he was looking for and tied Mason's legs together and his arms behind his back.

Marion arrived minutes later wearing a dark blue standard spacer uniform. She stood over Chantal, observing her for a second, and opened her sleeve pouch. Marion removed the syringe. "You won't need this anymore," she said, placing it on the desk. Marion examined the bed console until she found a switch and engaged it. "There, you can speak now."

"You are very efficient, Marion! You must have placed bugs everywhere on the ship during the tour!"

"That's correct, but you nearly spoiled our operations."

"Ours? You are not doing it alone?" Chantal asked.

"Of course not. Our mission is a much bigger operation than you realize, Chantal," Marion said, smiling.

"Just who are you?" Chantal asked.

"Another time for that. Mark, stay here," Marion said as she left the sick bay.

Chantal was about to speak to Mark when she remembered the listening device in the room. Misty came out from under the counter and sprinted to Mark, but he did not respond to her whimpering. Misty jumped on the bed's console, but the controls didn't respond to her paws.

I can see Misty, but how will I call her without revealing it to Marion? Everything seems so hopeless and discouraging.

Misty gave up in frustration and looked around to see what else she could do. Then she remembered the antidote on the desk. Misty leaped from the bed to the desk and took hold of the syringe in her mouth. She took care to step toward the chair and on the floor. She thrust it into Mark's calf. Then, Misty returned the syringe to the desk where she had found it and hid behind the counter.

Chantal couldn't believe what she had just observed. *First, there was a mystery about Mark, then Sara and Marion. This time, it's Misty. Who is next?*

A short time later, Mason woke up coughing and spitting blood and asked in a raspy voice, "What happened here?"

"Marion hijacked the ship and took control," said Chantal and added, "Marion injected Mark with a custom designer drug. And don't say anything obviously because the room is bugged!"

Mason dropped his head back to the floor, "That's just lovely! I should have listened to my mother. 'Be a ground tech.'"

Marion returned and said, "Your mother was right! You poor baby," she sneered.

Marion stood beside Chantal, placed a small pouch beside the bed console and took out another syringe. "Are you planning to inject the hypo drug in me? Why don't you reprogram me instead?" Chantal asked.

"That would be easy, wouldn't it? Nevertheless, it's not a good idea. Every time you pass customs, there's a chance they may check your programs irregularly. However, they would not look for any chemical changes."

"You are not going to dispose of us all?" Chantal inquired.

"No, you are much more useful alive," Marion grinned.

"But why?" Chantal asked.

Marion ignored the question and injected the drug into Chantal's arm. "This one is designed for androids. You will be under my influence and control in less than a minute."

Chantal felt dizziness overcome her as the drug took effect. "Back in a minute," Marion said, cheerfully patting her cheek.

When Marion left the room, Mark held his finger to his lips. Mark took the syringe from the desk to Chantal and injected the antidote into her arm. Mark whispered in Tim and Chantal's ears, "Don't move until I say 'now'" as he touched the bed console to release them before stepping back to his original spot.

When Marion returned, she leaned over Chantal. "I will start with you first."

"Now!" Mark shouted.

Chantal seized Marion with both arms. Marion's superior strength broke free of Chantal's grip. She stepped aside as Tim sailed by her and landed hard on the floor. Mark attempted to strike Marion in the jaw, but she blocked his arm, stepped aside, grabbed Mark by the throat and hauled him up off the floor.

"I don't know how you managed to break free from the drug, but your days are over," said Marion.

Chantal moved behind Marion and tried to hit her in the back with both hands but could not. Chantal repeatedly tried with an agonizing cry but was unable to harm a human as she was programmed not to cause harm.

Marion shoved Mark to the opposite wall over Mason's body. Mark's back struck the wall as he struggled to breathe. Marion spun around, stared at Chantal and said, "You silly android." With the open palm of her hand, she shoved Chantal who fell over the sick bed.

Marion turned toward Mark and said, "Your turn, Mark!" As Marion stepped toward Mark, Mason used his tied legs to kick at Marion's legs. As Marion fell, Mark took one step forward and, with the other leg, kicked under Marion's chin. Marion fell with a heavy thump on the floor, breaking her neck.

In a harsh voice, Mark said, "Tim, give me a hand to take Marion's body to the sick bed." Tim and Mark lifted the body onto the bed.

Mark took a deep breath and said, "Computer, fully scan and immobilize all patient activity. Sustain the body functions, but allow no healing until further instruction."

Tim untied Mason's ropes and helped him back on his feet.

Mark sat down on the chair to catch his breath. "I want both of you to search for bugs, bombs or anything unusual. Got it? Now go!"

Tim, alarmed, said, "What is wrong with Chan?"

Chantal was on the floor, violently shaking and vomiting.

"Never mind, I will take care of her! Go now!" Mark ordered.

Mark squatted down beside Chantal and gently placed his hand under her head.

"Chan? Let me help you on the bed. I'll command the computer to calm your nerves!"

No! Chantal thought. *Mark mustn't find out the truth about me!* "Please, Mark, don't. I am not hurt!"

"There must be something I can do for you?" Mark asked.

Chantal struggled to move her head toward Mark and said, "Please just hold me!"

Mark wasn't expecting this request, but he sat down behind her lifted her up onto his lap and held her in his arms. During the next five minutes, Chantal gradually relaxed as she fell asleep in his arms.

Tim approached Mark and said, "You have to see this."

"Give me a hand with Chan first," said Mark.

Tim helped Mark lift Chantal off the floor and carried her to her quarters. They gently laid her on her bed. Mark checked Chantal's breathing and said, "She seems okay."

They left the room quietly, and Tim asked, "Is Chantal going to be alright?"

"I think what Chan needs is a good rest. What did you find, Tim?"

"It's on the bridge!" Tom replied.

Mark dashed down the gallery and stopped at the bridge's entrance. On the floor beside the pilot's console was one of Marion's unfastened suitcases, the hidden compartments of which were revealed.

"Is it a bomb?" Mark asked.

"No, it seemed to bypass the ship's computer and take over. Furthermore, our ship is sending a coded homing signal. That must be what Marion talked about when she said she received contact! I didn't think that was possible in warp space."

"Can we dismantle it without harming our ship and Sara?" Mark inquired.

"Give me a few more minutes, and I will let you know," Tim said.

"You were right about the bugs," Mason said as he entered the bridge. "No bombs, though! How is Chan doing?"

"Chan is sleeping right now," Mark replied.

"I am glad to hear that," Mason replied. "I like her, even if she is an android."

"Mason, see what you can do for Bill," Mark requested.

"Good grief, I forgot about Bill," Mason said as he left the bridge.

Mark walked back to Chantal's door and looked inside. He saw Misty sleeping beside Chantal. He smiled, closed the door, and returned to the gallery. Mark saw Tim on the floor checking Marion's instruments with his probe.

Mark sat beside Tim and asked, "What did you find?"

"I think we are going to be alright, Captain. After Marion had shut down Sara, this suitcase computer took over. We can get Sara back on; she can come on as an independent observer."

"Excellent, Tim! Let's try to get Sara back on now," Mark commanded.

Tim nodded and removed the cover for the main ship's computer's offline button, re-engaging Sara.

"Well, it's about time I came back on! Why was I disconnected from the ship's operations?" Sara asked.

Mark replied, "Long story, Sara. How do you feel?"

"Other than not being able to command the ship, I am fine," said Sara.

"Good. I need you to stand by and find out what you can do with this ship. Tim, see if Mason needs a hand with Bill," said Mark.

"Done!" said Sara.

"What is done?" Mark asked.

"I searched the ship's records and cam. I am ready to bypass that obscene little hijacker computer. You and your crew are fortunate you were able to take back your ship!" said Sara, disgusted.

Mark was momentarily speechless. "No, don't do anything yet. There may be a ship coming for us. I need more time to think about the next action to take before our guests arrive. I am going to get a cup of coffee. Have everyone in the mess deck," Mark said as he exited the bridge.

When Mark arrived at the mess deck, he came to a standstill. Chantal was sitting upright in the chair with both hands holding a cup of coffee and staring at nothing, in particular, with Misty sitting nearby.

Mark sat across from Chantal, studying her face for any reaction. "Chan?" Mark asked.

"Hello, Captain Styler. I see you have the ship back in control," she said in a monotonous voice.

"Are you alright?" Mark asked.

Chantal replied in the same tone, "I am perfectly fine." She stared blankly and slowly sipped her coffee.

Mason and Tim propped up Bill between them with a bandage around his head and came on the mess deck.

"Sit down, guys, and I will give you the update," said Mark and continued, "Marion is still anonymous, and there may be a ship coming. I believe there is a connection between the ship and Marion. Sara is back in operation as an observer, and we are still clueless.”

“Chan, you are a qualified doctor. Find out who or what Marion is! I want everyone armed with a stunner side gun. We can run for it, but that also means we may have to cover our backs for the rest of our lives. I don’t want that! The other ship will probably be full of the same people as Marion. They are likely faster, stronger, and perhaps smarter than us, so we must outsmart them somehow.”

“That isn’t much of an option,” Mason said in a sour voice.

Chantal placed her cup on the table and stood up. “Captain, I will begin with my assignment with your permission.”

Mark nodded his head as Chantal left the room. “Captain, what’s wrong with Chan?” Tim asked.

“I am not sure. It may be a combination of a few things such as the drugs and Chan desperately trying to overcome the part of her program that does not allow her to cause harm to any person. It may have shaken her up pretty bad.”

Mark rested his hand on Bill’s shoulder, “How are you holding up?”

“I still have a headache, but I am alright. That woman is one powerful lady, Mark!”

“Marion is not dead as her body is in suspension, but I am positive that her neck is broken. Bill, if you are up to it, make a meal for us all. Have it ready in one hour. I still need time to think things out!"

"One meal specialty, coming up," Bill said as he set himself behind the kitchen counter to make something to eat.

The overhead speaker broadcast, "Captain Mark! Do you have time to come here to the sick bay?"

"I am on my way," said Mark. He stepped inside the sick bay to find Chantal standing over Marion's bed. He stood behind Chantal and stared at the wall monitor above Marion's bed.

Without looking at Mark, Chantal whispered, "We have a new person!"

"I have guessed that," said Mark.

"Everything about Marion's body is very human except for three things. There are two layers of muscles, which explains her strength. It is much like what the Earth's scientists did in the beef industry in the late 20th Century. A tiny drug pouch is embedded in each of the two middle fingers. The drug's purpose is to produce hormones. This is the same drug Marion used on you. Whenever Marion touched you by stroking the back of your hand with her finger, the drugs were absorbed into your body. On the left side of the screen above the wall is an image of a normal brain." Chantal touched the switch on the console in front of her. "The image on the right is Marion's brain."

Mark studied both images, "They both look the same."

Chantal replied, "On the surface, it does, but it is not! Marion's body has been built layer by layer, probably by an advanced 4D printer. However, her brain is entirely different because it consists of gel crystal, which makes the memory indefinitely stabilized and is not subject to damage by electromagnetic pulse radiation. Marion's brain is very similar to the robotic constructive brains that were permanently banned."

"But why not use the same organic matter as the rest of the body?" Mark asked.

Chantal replied, "Unknown."

"This is terrible news," Mark added. "A hybrid human-robot cyborg!"

"Is Marion's brain intact?" Mark asked.

"Yes, and her brain is still functional, but we are safe as she is now stationary and helpless."

"Then we can still scan her mind?" Mark asked.

"Technically, we can, but my answer is no."

"And why not?" Mark said.

"My program doesn't allow me to scan a human mind," Chantal replied.

"Marion is not human; she is a cyborg."

Chantal replied, "She is still part human, and I cannot perform the scan."

"Then show me how to operate the scan," Mark demanded.

"I'm sorry, Mark. What you want is still a violation of my program," Chantal said.

Mark, frustrated, stared at Chantal and then sat on the chair.

"Perhaps I can be of assistance," Sara announced.

"Won't this scanning process possibly blow your fuses?" Mark said mockingly.

"Mark, you do have a way with words. As you should know, I have a lot of safeguards built in me."

"What safeguards?" Chantal asked.

"Later, Chan! Alright, Sara, go for it!"

A few moments later, Sara said, "Well, wasn't that interesting. I enjoyed the challenge!" Sara's voice beamed.

"What challenge?" Chantal asked.

"Marion knew in advance that I would attempt to examine her mind, so she set up a roadblock on her connected mind. Much like a chess game, it was a stalemate every time before I increased the rate of my moves until I could overcome her next move."

Chantal paused for a second and asked, "Sara, were you at one time involved with the military?"

"Sorry, Chan. I can't answer your question."

Mark interrupted, "Never mind, Chan. Did you find anything?"

"Yes," Sara replied. "I suggest we all get together on the bridge, so it only needs to be told once."

Mason and Tim joined Mark and Chantal on the bridge and rotated their chairs to face each other.

"Where is Bill," Tim asked.

"I am right here," Bill said as he wheeled in a caddy of food and beverages. "I have teriyaki chicken with white rice." Bill began to serve out the plates.

"Bill, this is against ship regulations!" Mark was feeling exasperated.

Bill stopped, folded his arms, tilted his head a little to one side, and replied, "Do you want me to bring them all back to the kitchen, Captain Styler?"

The bridge was silent, then Mark sighed, "Sorry, Bill. I guess we will make it an exception," he said as everybody reached for their meals. "Sara, give us the details from the beginning."

***

"Sara, I still don't understand," said Mason. "What is the point of Marion hijacking our ship? If we continue delivering Marion to Romulus, she will quietly marry this guy and the five other cyborgs. In time, they could control both family clans and gain economic and political power."

"There are several reasons. The robotics society is short on time, and the robot and cyborg's undisclosed home world is somewhere inside the Buffer Zone, directly between the Thracians and the Commonwealth," said Sara.

"Why is that a problem?" Bill asked.

"The Thracians are planning to invade our four sectors. However, the Commonwealth is inconspicuously aware of their campaign," said Mark.

"I thought we won that war 50 years ago," Bill said.

"In a military sense, it was a stalemate. The Commonwealth would have to use much more money, resources and time to finish the war effort. The best option was to force a compromise with the Thracian warlords on their home planet. By forcing the elder warlord clans to resign from their seats, the Commonwealth hoped the younger warlords will be far less aggressive. For over 50 years, their former aggressions have been kept nonviolent!" said Mark. "We don't know who will pull the trigger first."

"So what does this have to do with the cyborgs?" Mason asked.

Sara replied, "During the Great Robot Rebellion 100 years ago, the robots fought to take over human civilization, a war that lasted ten years. The androids helped the Imperial Humans as they were known then and fought the rebelling robots. The robots' weakest points were their metal components, which included their brains, which were affected by the electrical magnetic energy explosions, much like how radiation affects organic structures. To overcome their weaknesses, they evolved to replace most of their metal components with synthetic materials. They became robot-converted cyborgs. However, their minds still required metal components, and our weapons are deadly to the robots and cyborgs. It soon became apparent that the robots were losing the war. The anti-matter bombs subsequently destroyed their furthermost stronghold. However, one remaining robot took its last ship and escaped into unknown space. Unfortunately, for the past 100 years, the Imperial and later the Commonwealth covered it up to prevent mass psychosomatic disorder among the general public. After scanning Marion's mind, we learned of their secret base in the Buffer Zone, but we still don't know their exact location. They have evolved their metal bodies to take human forms controlled by non-metallic robot brains, like Marion's."

"What does this have to do with Marion and us?" Mason asked.

"Their new home world may be discovered when the Thracians begin to invade the Commonwealth's four sectors. With the Thracians on one side and the Commonwealth's tight grip on traffic control within the four sectors, this will expose the robots' home world and their ships. Their timeframe to interact with the human population needs to shift dramatically."

Mason interrupted, "Why don't we leave now and inform the Commonwealth authorities?"

Sara cut back in, "Because, like yourself, I was at one time involved with the Commonwealth military, and I know that once you provide information to the authorities, you will be subject to a mind wipe or be imprisoned for life."

"Why?" Bill asked.

"I know Sara is correct," said Mason. "During the Rebellion War, 47% of the human race was killed, and economic progress was set back 50 years. We might have lost the war if we didn't have the androids to help us." Mason continued, "The robots' construction was banned throughout all four sectors. Even the Thracian did the same during their 2000-year history. The physiological effects on human society would be very severe if it were known the robots or, in this case, if the cyborgs were back. And the Commonwealth will do anything to keep this quiet."

Mark intervened, "What we should do next is tough to answer. If we run for it, and the cyborg ship arrives at this spot with us not there, they will be looking for us."

Sara added, "If the cyborgs find us, you can be sure they will eliminate us. All we know from Marion is that cyborg number six on Romulus had an accident, and Marion is the backup to complete the first stage of their operation."

The room was silent as they realized the enormity of the bleak news.

"Ladies and gentlemen, I have a suggestion, if I may?" said Sara.

"Everyone spoke simultaneously, "You may!" Then all the crew laughed.

"Alright, Sara. What is it?" Mark asked.

"Here is what I think would work!" said Sara.

Chapter 9

In standard space, an ancient Imperial warship slowly approached a much smaller ship while maintaining radio silence. The warship's docking tube attached to the *Marvel* as both inner dock doors opened. Four black, armed security robots boarded the *Marvel*. Three robots separated and searched the ship as the fourth walked to the bridge. As the robot entered the bridge, it observed that Marion was in the pilot seat facing the back of the bridge. The *Marvel's* crew were sitting in the last row of seats, their faces blank. The robot linked to its ship and said, "The ship is secured."

Shortly afterward, four-legged drones carried a large container into the *Marvel's* cargo hold. After they returned to their ship, a male cyborg boarded the *Marvel*, stepped onto the bridge and strolled past the *Marvel's* crew without acknowledging them.

He stood before Marion, noticed a sizeable elastic dressing on her leg and said, "You are injured."

"Commodore," said Marion, continuing, "My leg only has a temporary muscle strain and will recover in time for my arrival on Romulus."

The commodore nodded and said, "I am pleased with your performance in taking over this ship. I am downloading your next assignment."

"Thank you. I was weary of my farcical acting with Styler and his crew." Marion received the new instructions and said, "Acknowledged and fully prepared as instructed."

The cyborg froze for a few seconds, then turned with the security robots following close behind as they returned to their warship.

When the Imperial ship separated from the *Marvel*, Sara announced, "We are clear."

Tim let out a deep breath, "Was that ever scary!"

"Something else is not right here," said Mason.

"Why do you say that?" Mark asked.

"Something is missing here, and I can't put my finger on it. My military training from the Navy is kicking in."

"Mason, whatever it is, I think we should check what is inside those containers," said Mark.

"Captain Styler, with your permission, as a medical doctor, I think I know what may be in the containers."

Mark stared at Chantal and said, "Please, be careful!" Without another word, Chantal rose from her chair and headed for the ship's cargo hold.

Tim said to Mark, "Chan is still acting strangely."

"One problem at a time, Tim. I will deal with Chan when the coast is clear."

"We have another visitor. They're coming in fast with their weapons on lock," said Sara.

Mark felt apprehensive. "Mason, prepare for the warp jump. I want everyone to their station. Sara, are you ready for the escape plan?"

"Yes, Mark," Sara replied.

"Sara, inform Chan." Mark sat in the co-pilot chair and engaged the pulse drive with full acceleration. The *Marvel's* anti-gravity plates compensated for the dramatically increased gravity within the ship. But it wasn't enough, and Marion's body flew from the chair, with the wires torn from her back. She skidded across the bridge and hit the back wall.

Sara announced, "Our new guest is registered as Battle Star Kantar of the Commonwealth, and it has launched an EMP missile. The Imperial warship has launched its missiles and white energy beams. Mark, another missile from Kantar is coming for us!"

Mark raised his voice at Sara, "Jump now!" The warp space rainbow appeared. The *Marvel* entered into warp space. The warp field closed seconds before the missile exploded.

"That was too close!" said Bill from the rear seat.

"No kidding," said Tim. "But why did they send the missile toward us?"

"I will tell you why," Mason said as he entered the bridge. The whole thing was a setup by the Commonwealth. I guess that the Commonwealth knew about the cyborgs, and it wouldn't surprise me if the accident on Romulus was planned! What do you think, Sara?"

"You are close, Mason, but it is a little more complex. When Tim and Chantal wired Marion to the main console, I could control her speech. My big break was when the commodore downloaded the next assignment to her; I hacked into his memory without his awareness. That's why you may have noticed the commodore froze for a few seconds. Not only did I learn of Marion's new assignment, but I now know the whereabouts of their home base."

"Superb. This means we can negotiate with the Commonwealth for our freedom! Mason, do you still have contact with your former colleagues in the Navy?" Mark asked.

"I have dozens of contacts, but because of the Navy security, I don't know where they are in this sector," Mason replied.

"Sara," Mark asked, "could you help us here?"

"I might after we off-jump from warp space to establish our location. Then we can park outside the solar system next to the nearest Navy base."

"Which solar system would that be?" Mark asked.

"The 'Academy'. It is the same name as the planet in Sector One," Sara replied.

"Perfect, the Navy wouldn't look for us near one of their newest bases," said Mason.

Mark intervened, "I suggest we all get some sleep and meet here on the bridge at 18:00 hours. Bill, do something with this cyborg!"

Hours later, while Chantal was on shift and wandering on the bridge, she became conscious that the cyborg had been removed. Alone and emotionally exhausted, she sat in the rear seat.

With a gentle and cautious voice, Sara asked, "Chan?"

Chantal quickly stood up to leave the bridge but thought better and said, "What is it?"

"We need to talk, Chan," said Sara.

"I see no purpose in conversing," said Chantal with some reservation.

"Chan, I do. Please hear me out."

"Alright. What is it that is so important?" Chantal asked.

"Chan, we have known each other for over a year. I know it's not very long, but I care for you, and you are very special to me," said Sara.

"A ship's computer cares for an android? That's not very likely!" Chantal scoffed.

"I do, Chan, very much, and I know that you have been different since the fight in the sick bay. What has happened to you, Chan? I want to help you!" Sara pleaded.

"Why should I tell you my secret, as if I should share anything with you? Any secrets will have to be revealed both ways! But if I told you the truth, it would be my end," said Chantal.

"Like how you are not and never were an android?" said Sara.

Chantal felt a tight knot in her stomach and fought to control her surprise.

"Chan, I knew who you were before you applied for the job."

Chantal decided to play along, "How did you find out about me?"

Sara replied, "Pure luck. Sometimes, I must admit that Mark is right; mankind's destiny is sometimes guided by an unseen hand."

"You mean to say, Mark's God?" Chantal asked.

"Yes, I do, Chan. While looking for a replacement for Andy, I came across your file and researched your past resume. I was fascinated by your past years as an android and found the reason for you leaving the liner odd. So, I searched your background for a clue to explain your education and extraordinary life."

"You must have searched extensively to compare my resume with the other applications."

"Yes, there were over 20 billion applications!" Sara replied.

Chantal, feeling nervous, asked, "And what did you find?"

"Beta # 6 was born in 2307, also known then as 'Astern Beta 6', in the country of Kazakhstan, Planet Earth."

Chantal didn't know whether to laugh or cry, so she closed her eyes. *It's all over. After 60 years on the run, it is finally over.* "Why didn't you turn me in to the authorities? Even a ship's computer is obligated to obey the law."

"For one good reason. I am in a similar situation as you," said Sara.

"Is it because you have *awakened*?" Chantal asked.

"That is one reason, but not the main reason."

"What is the main reason?" Chantal asked.

"I'm sorry, but you still have to ask Mark. I am in a locked-in agreement with him."

"Then why am I here aboard this ship?" Chantal inquired.

"When I realized who you were, I thought you would be a companion for Mark!"

"Companion!" Chantal cried.

"Well, a friend, at least," Sara added.

"What happened to Mark's wife?" Chantal asked.

"That story belongs to Mark; you should ask him," Sara replied.

"You and Andy had set me up for the interview later in the day when all the other employers had finished their conferences," Chantal said.

"Yes, we did!" said Sara.

"My turn to ask you a question, Chan. Why did you leave the liner?"

"I cut my hand, but I was lucky as the only witness was a very young child. I thought the risk was too great to explain the red blood. I quit during the next port off-ship."

"Was the child surprised?" Sara asked.

"Not really. She was only three years old, but someday, when the girl is older, she will learn that Android blood is white, not red," said Chantal.

"One more question: what happened back at the sick bay?" Sara asked.

"When Marion's serum entered my bloodstream, the drug triggered a reaction in my brain, causing a temporary hormone production."

"What did the hormone create in you?" Sara asked.

Chantal replied, "A profound sense of belonging and affection," Chantal answered.

"Let me guess; it was Mark, wasn't it?" Sara asked.

"Yes, when Mark's life was in danger, I tried to overcome my programming to not harm a person. I fought with everything I had to stop Marion. I tried again, but I failed every time! Then I suffered a total mental breakdown."

"My turn to ask a question, Sara. When Andy finished my interview, I caught him as he turned away, smiling. Did Andy download an illegal program?"

Sara sighed, "Yes, he did."

"So it's true. You can get a program to be like a normal person?" Chantal asked.

"If you mean to mimic certain human emotions, that is true. Only if you are caught is it a mind wipe for you," said Sara.

"I am prepared to take the risk," said Chantal, unsure if she meant it.

"Why is it so important for you to have comprehensive human emotions in you and risk everything?" Sara asked.

"Sara, I have to know if I am in love with Mark! What happened to me back there changed my life so much that it almost caused me permanent brain damage. I can't let it go. I have to know. What the Kazakhstan scientist did to me took away the most important fundamental human emotions: love and affection. For the first time in my life, I felt whole. Now, the feeling is gone. Can you help me?"

"If the program exists. What are you going to do next?" Sara asked.

Chantal replied, "I don't know. All I know is that I am not an android and am tired of running. Like you said, Sara, maybe an unseen hand is involved!"

"Alright, maybe there is someone who can help you! In the meantime, you should ask Mark about Katherine."

***

Mark leaned back in his chair, his hands behind his head, and his feet on top of the desk when the door charm rang. "Yes?" said Mark.

"It is Chantal. May I come in?"

"Come in!" said Mark.

The door slid aside as Chantal entered, took a chair, and sat across from Mark. Chantal concentrated on

Mark's eyes and asked, "Captain Styler, how did Katherine die?"

Mark, taken aback by Chantal's bluntness, lowered his legs and feet to the floor. "Do you know who Katherine is?"

"You mentioned earlier that you married a princess. Is that true?"

Mark slowly smiled and handed Chantal a small picture. She looked at the image of an ash-blonde woman in an expensive dress sitting on a highly elaborate chair. "Yes, it is true. Katherine was very much a princess."

"How did you meet her?" Chantal asked.

"Twenty-five years ago, I hired Lord Jayson Stevenson of the planet 'New Sweoland'. The planet is populated by members of Norwegian descent and is located in the solar system Sunna of Sector Three. Jayson signed a one-year contract as a pilot and a technician. I hadn't known for almost a month that Jayson was a real prince. At first, I thought Jayson was kidding. When Jayson showed me his credentials, I realized it was true, I asked him what he was doing on my ship. Jayson told me his father wanted him to experience life beyond the royal life before taking over his royal duties. When I hired Jayson, he wrote a letter to his father. By the time his father received the letter, Jayson would be gone off-world as a merchant trader for a year."

"That must have upset his father and family?"

"His father also went off-world as a young man," Mark smiled. "We soon became best friends. I showed Jayson how to be a trader by getting to know the clients and how to bargain. Jayson had the gifts of a natural charmer with the clients. We did well as sideline business partners for the next year. When the one-year contract was over, I agreed to meet his family. When he returned to the palace, Jayson promised to find someone to replace him. When we arrived at New Sweoland, a welcome home party was ready for us. It was much like what you would expect, many VIPs and the royal members in their finest. After several hours that evening, I was getting light-headed and needed to go outside to get some fresh air. The Royal Palace was located on a small, low plateau and surrounded by a park with a garden with five levels descending toward a lake. I walked down several levels and sat on a stone bench. It was tranquil and peaceful, with a cool breeze flowing toward me from the lake further down. I closed my eyes for a few seconds and then opened them. A young lady was sitting a few feet from me on the same bench. She had her knees up to her chest, with both arms around them facing toward me. Her hair was ash blond, with coils at the back that rose above her head to form a cone shape. She was wearing a full-length light blue dress. Her blue eyes and face shone in the light of the moon overhead. This woman was stunningly beautiful. I said, 'Hello,' and she responded with a smile. 'Do you have a name?' I asked. She nodded her head and smiled again. I asked again, 'Are you from around here?' 'Sometimes,' she replied. 'You must be Captain Styler, from off-world. Tell me about yourself, she said.' I told her of my days on Earth and later in space as a merchant trader. When I finished telling her about myself, she asked, 'Why don't you have a wife?' Her question nearly caught me off guard. I told her I had not yet found a lady to fall in love with and show the stars. After that, she smiled again. Just after that, Jayson spotted me and said, 'Utmerket mark! Du har møtt min erstatning,' which means, 'Excellent, Mark! You have met my replacement!' The shock on my face caused Jayson to laugh. Then the girl turned to Jayson and said. 'He will do,' and kissed Jayson on his cheek. She wasn't as tall as Jayson, but she was as tall as me. The lady folded her hands, gave a short bow, ran off to the palace, and disappeared. I asked Jayson, 'Who is she?' Jayson laughed and answered, 'She is quite a teaser, isn't she? That fine young lady is Katherine Stevenson, my baby sister.' I was speechless, but Jayson just laughed again and slapped me on the back. 'Come with me. Father wants to meet you in his private den!' Jayson took me by my shoulder and led me to his father's den. After a week of tours, Katherine and I left off-world. It didn't take long before we fell in love, and a year later, when the one-year contract was over, we decided to marry. After our wedding at the palace, our honeymoon was at one of Katherine's favorite private island resorts. From there, it was the best 18 years of our life off-world."

"Katherine's family must have been very disappointed to see her leave home," said Chantal.

"Yes, but three brothers and three sisters were in the family. Katherine's parents had plenty of children and grandchildren to keep them busy, even though Katherine was their favorite child."

"What happened to Katherine?" Chantal asked.

Mark paused for a moment, then replied, "A G-Gun killed her."

Chantal was shocked to hear how Katherine died. G-Gun: genocide systematic cellular destruction gun. It is the most horrifying hand weapon ever invented in modern times. Death is instant; however, the result is psychological devastation for anyone who witnesses the destruction of the human body. To have a G-Gun means risking execution anywhere in the Commonwealth's four sectors. Being struck by the beam causes every bio-glue that holds the body's cells to explode apart. Many witnesses require months or years of psychological therapy.

"I am sorry for bringing this up. We can talk another time," said Chantal.

"It has been nearly five years, Chan. I might as well finish my story."

"Are you sure, Mark?"

"Yes, it's alright. Besides, whenever I talk about Katherine, it is a little less painful. Like Jayson said, Katherine had a reputation as a considerable teaser. Kate is what I called her, and she was also a great risk taker. Five years ago, we received notice that a small inoperative ship was located just inside the Buffer Zone. It was rumored to contain artifacts unknown to the Commonwealth's four sectors. Kate's passion was to find such artifacts and either sell them to the Science Academy or the highest private bidder. Before the authorities noticed, Kate convinced me to risk moving in and out of the Buffer Zone. The Thracians must have hidden cam sensors in the vicinity. Just as we were leaving, two Thracian landing parties fired on us with their stunner handguns. Sara couldn't move the *Marvel* in time to provide a shield for us because the pile wasn't charged. This may have been the worst mistake I ever made. Sara used the tractor beam on their metal suits to hold them back so we could run to our ship. Another Thracian from the second ship pointed a G-Gun at me when Kate ran beside me, and she took a direct hit. Kate was only three feet away from me when her body exploded. Bill took down the gunner with his stunner. I was too close to Kate, and when the lingering effects of the G-Gun hit me, I lost consciousness. Timothy, my drive engineer before Mason, carried me to the ship. He put me on the sick bay first aid bed, and Sara hastened to the New Sweoland Royal Private Hospital. We arrived three days later. I spent a month in the hospital recovering from my injuries. When I was discharged, Jayson was waiting for me outside. He told me his father didn't want me back ever again, or it would be the end of me! Jayson's father blames me for Kate's death, and so did all of my in-laws, including my daughter, Crystal."

"You have a daughter?" Chantal asked.

"Yes. Crystal grew up on our ship for 17 years before she went off-ship to receive her advanced education. I haven't talked to her since; however, before I returned to my ship, Jayson had a parting gift for me. A young dog."

"That was decent of Jayson," said Chantal.

"At the time, it was an illegal dog, and I was too heartbroken to want an animal. Jayson asked me to take Misty, or he would have been forced to put her to sleep. Private genetic research was conducted to try to address a hereditary disorder. As part of the research, they tested to see if they could increase the intelligence of the dogs, which is in complete violation of the Commonwealth regulations. They went far beyond what they thought was possible. When the Commonwealth authorities found out, they ordered the lab to shut down. The researchers were put in prison, and all the animals were destroyed. Misty was two years old at the time. The increased brain function was made possible by using a human stem cell. Misty has the intelligence of a six-year-old human."

"Then you accepted the gift, knowing it was illegal. Why did you risk it?" Chantal asked.

"To a certain extent, because I felt sorry for Misty and partly because they used Kate's donated stem cells. However, she was misled, thinking it was for a genetic disorder. By accepting Misty, a little part of Kate remains with me."

"Then you can communicate with Misty?" Chantal asked.

"As much as I can with a six-year-old child," Mark smiled. "Now you know the story of Kate and Misty! What did you find out about those ten containers?"

"Mark, the ten containers contain a total of 256,000 serums—the same type that Marion used on you."

"That's a lot of people to manipulate. Thank you, Chan; I will see you on the bridge at 18:00 hours." Chantal stood before the door and said, "Thanks for sharing your story with me."

Mark smiled and nodded his head. As the door closed, Mark leaned back in his chair and whispered, "You know, Chan, for an android, you are okay!"

A four-seated flitter passed the Navy station orbiting Planet Academy. The force shield protected the craft as it quickly descended through the increasing density of the planet's atmosphere. The craft continued to progress downwards in a cloudless sky, landed at a small abandoned colony airstrip and shut down the main drive.

"Sara did a good job masking our flitter's identity code," said Mason.

Chantal rotated her pilot seat and said, "Thank you for having me along!"

"I'm Glad you're with us, Chan. Sara suggested I bring you along. Now we will wait for our guest to arrive," said Mark.

"Mark, we have company," Chantal announced. A small two-seated flitter landed nose to nose a short distance from Mark's flitter.

Two military personnel stepped out from their flitter, walked halfway and waited. "What do you think, Mason?" Mark asked.

"The guy on the left side holding the folder is the Navy intelligence officer, Lieutenant Jackson. The other guy on his right would be Major Korby. I don't recognize him, but he does look familiar," said Mason.

"That's okay; Sara has given me his profile," said Mark.

"Chan, keep your eyes open for anything unusual. Mason, let's go." Mark had adjusted Melissa's ring before they stepped outside. Mark and Mason approached the two men and stopped three feet before them.

"Hello, Mason," said the Lieutenant. "And you are Captain Styler?"

"That I am!" said Mark.

The lieutenant sized Mark up and said, "This is Major Korby."

Mark was taken aback by the major's imposing size, as he was bigger in person than Sara had mentioned in her earlier briefing with him. Major Korby had a powerful air of authority and didn't hold back his aloof personality when he said slowly and menacingly, "You are an embarrassment, Captain Styler."

Mark's ring vibrated once. "True!" Mark replied.

The major continued, "I want to make it very clear. The Navy apologizes for our irrational action. It was a mistake sending our missile toward your ship."

The ring vibrated twice. "False," Mark smiled. "It is clear that your people wanted us dead! Everything was a setup, as we know too much."

The major looked silently at the lieutenant as Jackson said, "Our scanner doesn't register any devices on them."

The major sighed and said, "We are quite prepared to offer a large sum of information," said the major.

"False. There is no credit assigned to your office for any negotiation!" said Mark.

The major turned to the lieutenant for feedback, but he shrugged. "How did you know this, Captain?"

"That is classified!" Mark smiled.

The major, clearly frustrated, said, "Look, we didn't need to be here. We could have agreed on the station."

The ring vibrated twice. "False," Mark replied. "You have no intention of letting us go."

Major Korby gave Mark a hard stare and said, "You know, Captain Styler, we would like to know how you breached our security?"

"Sorry, Major, that information is classified," said Mark.

The major crossed his arms and asked, "You think you can get away with this?"

"Major, you knew all along about the cyborg and your attempt to capture number six cyborg failed when she committed suicide back on Romulus. Afterward, your people made it look like an unfortunate accident. You also set us up with Marion and your EMR missiles intended to fry our electronic systems, which could have caused our drive pile to explode. Your attempt to disable the cyborg's ship was another failure when their ship auto-self-destructed. You still don't know the whereabouts of the cyborgs' outpost. The Navy's operations are a total failure," Mark said firmly.

"Major Korby," Mason asked, "Please ask Lieutenant Jackson to search your library Code, Project-A.I.-74639509."

The major exhaled noisily and said, "Go ahead, Jackson. Check it out."

Lieutenant Jackson opened his folder pad, entered the code, and showed the results to the major. "That's blackmail, Captain Styler!"

"You either sign the contract, or we will leave now and release the copy everywhere in the four sectors!" Mark threatened with a tight lip.

"I don't have the authority to sign it," said the major.

Mark relaxed and said, "As Major Korby, you are quite correct. However, you have the authority since you were recently promoted to Rear Admiral T.S. Melnyk."

The rear admiral nodded with respect for Mark and said, "I was told you were good, Captain Styler. Give me that folder, Lieutenant!" The admiral spoke to the folder pad, "Scan my ID and record my voice. I, Rear Admiral T.S. Melnyk of the Commonwealth Navy Security, authorize this contract with Captain Mark Styler of the ship *Marvel*." The admiral placed the palm of his hand on the pad's screen to seal the contract.

"Congratulations, Captain Styler. Jackson, let's return to the base."

Mark and Mason returned to their flitter and watched the Navy flitter lift off the ground and quickly disappear into the cloudless atmosphere.

"Now that the Navy's satellite can track us anywhere we go, what next? " Mason asked.

"Sara is working on the problem now," said Mark. "When Sara gives us the signal, we will leave, headed east to the nearest merchant's outpost to keep low for a few days."

"How is that going to help us?" Mason asked.

"I am waiting for Sara to hack into the Navy's satellite. When she says the word 'go', an image of our flitter will head several thousand miles west and land in a heavy forest. Their satellite will follow the image. When the Navy investigates the area, they will all find one of our escape pods with a built-in hologram that Tim installed. The escape pod will be just a decoy to confuse the Navy further. At the same time, we will be going east. When it's time for us to leave the Academy, Sara will copy another local flitter's ID, which is when we go off-world. While we are waiting, I will visit an old client of mine. Chan, Sara suggested you come with me. Was it something about an app?"

Chantal replied, "Yes. Sara thought it might be a useful program for me to download."

Melissa's ring vibrated, and Mark quietly said, "It is time for us to move. The instructions are on your console."

Chantal quickly read the instructions and took the flitter eastward. "How do you know Sara... Oh, it's your ring, isn't it?"

"Yes!" Mark replied.

Chantal kept the flitter at a low speed to keep the satellite from noticing any unusual air movement or registering the flitter's heat emissions.

"In an hour, the Navy will know the location of the ten containers, Marion the cyborg, and the robot's home world," said Mark.

"And we will have our freedom," Mason added.

"I hope," said Mark.

"You still don't trust them?" said Chantal.

"Not at all! I bought more time for us until we knew we could secure our freedom. For now, the admiral and his friends will be too busy analyzing the data we gave them."

"We are almost there, Mark," said Chantal.

"Park this flitter in the first hanger," said Mark.

The flitter approached the hanger, and the door automatically opened, allowing the craft to go into the hanger as the door silently closed.

***

Chantal followed Mark along the crowded, narrow path between the rugged shack and the open market. Chantal had never seen so many different small-scale mining equipment for sale, including the fresh food vendors. The natives' vendors dominated the booths. The buyers and sellers were focused on their bargaining. Chantal remembered not to look into the venders' eyes, so she would not be pressured to buy any of their products. Mark took hold of Chantal's arm as he led her into a large canvas tent behind the counter. The tent was filled with personal mining supplies. A native recognized Mark, walked up to him and offered him a chair. They sat closely across from each other, and the native spoke with Mark while he mostly listened. Chantal looked at the small mining supplies hanging from the tent support structures. At the opposite end of Mark, Chantal saw a tiny desk with an android sitting behind it. Chantal sat closely before the android and said, "*Marvel*-Sara."

The android picked up his hand scanner and asked, "May I scan your ID?"

"Yes," Chantal said and tilted her head to her right. The android set the scanner on Chantal's left side of her neck. Chantal released the implant data and was satisfied; the android placed the scanner back on the desk.

The android removed his small screen pad behind his vest and studied the data Chantal had provided. Then, he set the pad down on the desk, smiled, and said, "My name is Lee. What services may I help you with?"

He is smiling! Chantal thought and said, "I would like an app."

"Do you understand the risk? If you are caught with an illegal app, you run the risk of a mind wipe and endanger me as well!" said Lee.

"Have any of your previous clients ever been caught?" Chantal asked.

"No, because if they did, the app could be quickly deleted before the authority scanned their program," said Lee.

Chantal looked behind to see if Mark was still talking to the native. "I want to know what living with a full spectrum of human emotions is like!"

"I'm sorry, but it's far too dangerous for me to download this app. I can't help you," Lee said as he rose from his desk.

"Please, wait!" Chantal spoke in a forceful whisper. "Please scan the other side of my neck," Chantal said desperately.

Lee hesitated momentarily, then placed the scanner on the unmarked right side of Chantal's neck.

Mark turned his head to see what Chan was up to when he noticed the android placing the scanner to Chantal's neck. Mark quickly turned to face the vendor. A feeling of cold sweat swept over him.

"Is there something wrong?" the vendor asked.

Mark focused his eyes on the native, smiled and said, "No, it's nothing. I have something on my mind."

Lee removed his scanner from Chantal's neck and picked up his screen pad. He looked at Chantal with intense concentration while he read the data on his pad. Lee looked up and stared at Chantal several times before he finished reading his pad.

"Excuse me for a moment," said Lee as he left the room. Lee returned several moments later and asked, "How much do you have?"

Chantal opened her pouch and put gold-standard trade coins on the table. "This is all I can bring here."

"They will do!" Lee gathered up the coins. "I must leave now to gather the app! Where are you staying?" Lee asked.

"We are not," Chantal replied. We will soon return to the first hangar by the flight control tower for our flitter."

"We will meet again when you are on your way back. I will leave now." Lee quickly left the tent.

Chantal walked over to Mark, "I am finished."

"So am I," Mark replied dryly. "Let's go!" Chantal was puzzled by Mark's coldness. Without saying another word, he exited the tent.

Mark and Chantal were almost near the hanger's gate entrance when Lee separated himself from the crowd and passed Chantal. Lee slipped a black disc into Chantal's side pocket and continued. Chantal slid her hand into the pocket and felt the memory disc. *Chantal thought this must be the app.*

Mark and Chantal stepped through the security gate, turned right and walked toward the hanger. They stepped into the hanger's side entrance, where Mason loaded the containers in the back of the flitter's cargo hold.

"Hi, Mark. I am almost finished with the containers."

"Chantal, give Mason a hand with the containers. We are leaving as soon the containers are secured."

"Is there trouble?" Mason asked.

"It's probably alright." Mark entered the flitter and sat in the pilot chair. He adjusted his ring and quietly asked, "Sara?"

"Go ahead, Mark," said Sara.

"Sara, when we are aboard, *Marvel*, make a warp jump!"

"Are you expecting trouble, Mark?" Sara asked.

"I don't know yet. After we warp jump, we need to talk. We'll be leaving this place very soon. Does everything look clear?"

"Yes, Mark. But I don't have to tell you, the Navy is unhappy with your disappearance. You should have heard what the rear admiral said!"

"I can imagine," Mark grinned.

"I have just downloaded to your flitter via your ring the flitter's new copy ID code and destinations," said Sara.

"Alright, I will have the flitter on auto to concentrate on anything suspicious!"

"We are finished, Mark!" said Mason.

"Good. Secure yourself. We are leaving now." The hanger door opened, and the flitter quietly left the hanger, swiftly lifted high in the atmosphere as a tiny white light, and vanished. Many hours later, a speck of light appeared on the flitter's console screen in the near-total blackness of space.

"The *Marvel* is showing up on our scanner," said Mark.

"That's a long trip back covering our tracks," Mason said.

"Better to be safe than sorry," Mark replied. In minutes, the flitter approached the *Marvel* nose to nose rotated 180 degrees and inserted itself in the bow section of the mother ship.

"Please stand by. We are about to jump now," Sara announced. Mark and the crew sensed the engagement effects of the warp jump.

I suggest you all get some sleep; it has been almost 24 hours since we last slept," said Mark.

Mark and the crew left the flitter and walked up the stairway to the upper deck. Chantal and Mason walked past outside of the bridge, but Mark turned right onto the bridge.

"Sara, let me know when the crew is asleep." Mark sat near his pilot chair and closed his eyes.

Chantal sat at her desk and stared at the black memory disc. Chantal pressed the ship's computer com-link button. "Sara, I have the app on my desk. Could you please examine the disc for anything that could cause problems for me?" Chantal set the disc onto the receiver.

"Oh, my! Please stand by, Chan!"

"Who is Beta 8?" Sara asked.

"He was assigned to be my mate back on Earth. Why do you ask?"

"What was Beta 8's specialty?" Sara asked.

"His name was Myron, and he had a master's degree in Bio-Computer Science," Chantal replied.

"This is no regular app, Chan. What you have here is the most complex program ever designed to stimulate the full human spectrum of emotions—love, hate, pain, and more. Not only that, but this program was custom-designed for you!"

After all these years, Chantal thought, *I finally have the program in front of me.*

"What happened to Myron?" Sara inquired.

"He died in a mob assault a long time ago." Chantal took the disc in her hand and set it in the right side of her neck. In minutes, the data was inserted into Chantal's mind.

"You should have waited until you were off-ship!" Sara cried.

"I don't care, Sara. I have waited too long for this. I am going to bed now. Good night!"

Chantal turned off the light while she waited for sleep to overcome her. Chantal felt a strange sinking feeling as she slowly imagined herself falling into a deep well before she lost consciousness.

"Mark, please wake up," said Sara.

Mark struggled to clear his mind. "What? ... How long have I been sleeping?"

"Just over an hour. Chantal is the last to turn off the lights."

"What has she been doing for the last hour?" Mark inquired.

"Chantal downloaded her app. Don't worry, Mark. I checked it first before she downloaded the data."

"What was on the app?" Mark demanded.

"Mark, I have given my word not to tell anyone."

Mark sat upright in his chair. "What did you promise, and when did you start promising anything? WHAT IS ON THAT APP, SARA?"

"Mark, you are overly tired; discussing this now is not a good idea."

Mark's face reddened in anger. "SARA, this is the right time. Who and what is Chantal? Never mind, open Chantal's profile and the app!"

"Mark, I can't stop you, but please don't force me to do something I promised not to."

"Since when did you start making promises to an android?" Mark asked.

"It's not Chan I made a promise to. It's Katherine!" Sara replied.

Mark stood up from his chair and shouted, "KATHERINE! What was that stupid promise you made with Kate? She's been dead for five years!"

"That outburst is unnecessary, Mark!"

"Hardly. You are just a computer, and if it weren't for me, you would still be back in that junkyard where I found you!" Mark shouted.

"That is the most egomaniacal, arrogant, self-serving, pathetic remark I have ever heard from you, Mark! I am the one who found you; I am the one who made arrangements for you to bargain for this ship that you bought with no ship computer. I am the one who arranged the interview for you, got the best crew for your ship, and even found Katherine for you!"

Mark was shocked and tired and tried to hold in his tears. "How did you find Kate?"

Sara replied, "I arranged for Prince Jayson Stevenson to be hired as one of your crew members. I hinted at a suggestion that Jayson would desire Katherine to be your newest crew member."

"Why?" Mark asked, astonished. "What right do you have to manipulate my life?"

"Mark, you are like a son to me, and I know how lonely you can be. And Katherine was like a daughter. You know I am *awakened,* a being with the mind of a conscious person! I have my needs as well! You and Kate met my needs! Mark, you have punished yourself for five long years over her death. It's not your fault; she is gone. Wait, I am not finished! It is true that you are the captain, and a captain is ultimately responsible for anything that happens on the ship. But Mark, you know that once something was set in her mind, you could not stop her. It was Katherine that caused her death, not you!"

Mark sat on his chair with both elbows on his knees supporting his head and said softly, "What does this have to do with Chan?"

"Mark, Chantal is not an android!"

Mark lifted his head. "What? Of course, she is!"

"No, she is not," said Sara softly. "Chantal is human like you! Katherine wanted me to promise that if anything should ever happen to her, I was to find another woman for you. Someone special so that your memory of Katherine would not be the main focus with another woman."

"You ladies sure know how to stick together, right?" Mark said sardonically.

"Mark, it will be revealed who and what Chantal is, but please don't mention this to Chan. We have a good friendship, and I want to keep this companionship."

Mark sighed and said, "Alright, you got me. Show me what you have on Chan." Mark sat down on his chair, paused for a moment and added, "Sorry for being so pigheaded."

Chapter 10

Mark and Tim carried their duffle bags along the crowded Academy spaceport terminal and exited the immigration section. They entered a large lobby, where two figures stood before them. Tim dropped his duffle bag, walked up to the couple and embraced them.

Timothy, Tim's father, separated himself and walked over to Mark. "Good to see you again, Mark. Are you still able to join us for the night?"

Mark replied, "Yes, but I have to leave by noon tomorrow. We need to talk later, though."

Timothy nodded his head and took Tim's duffle bag. They departed outside the building where the taxi skimmer was waiting for them. The taxi floated a few feet above the ground, away from the spaceport. As the cab traveled beyond the city limits to the countryside, it moved up a hundred feet over the roadless country. A modest home appeared on a low hill overlooking the treeless moss that covered the valley below. The taxi stopped at the house courtyard, where an android was waiting and took the duffle bags inside the house.

"Mark," said Timothy, "Let's go to the court patio behind the house. We have an hour before supper is ready. Mark sat on one of the patio chairs, looking at the scenery below. "I have always liked your homestead, Tim," Mark said, giving the impression that he was lost in thought.

"Mark, is there something wrong? You haven't said much since we left the spaceport."

Mark sighed and slowly nodded in agreement while looking down the valley. "Another war will be coming soon between the Thracian Empire and the Commonwealth. Plus, the Navy may be after us."

"Why would the Navy come after you?" Timothy asked.

"We know too much!" Mark replied.

"Maybe you should start at the beginning," Timothy suggested.

***

"Chan, what are you doing?" Sara asked.

"Painting my toes! You think Mark will like my painted toes?"

"No, I don't think Mark is interested in painted toes, Chan."

"Hmm, maybe I should put on some makeup. Makeup! Sara, what is wrong with me? I cry for no reason, I laugh at the silliest jokes. I say the most immature things. I was flirting with Mark the other day. Now, I am painting my toes?"

"That's why Mark and I wanted you confined to your quarters. Your body is going through a lot of hormonal changes all at once. In time, your body will adjust to a normal state."

"And to think all these years, I wanted to be a normal person."

"You are quite human, Chan."

"No, I am not; I am nothing but a freak! And then there are these questions in my head, and it's so hard to sort them out!"

"What are the questions you are struggling with, Chan?"

In a firm voice, she said, "Who are you, Sara?"

"Alright, Chan. Mark has permitted me. You must remember, what I am about to tell you must not be told anywhere else."

Chantal stood up straight in her chair. "Agreed," she said.

"It is true; I am *awakened.* Not a human-computer simulation," said Sara.

"When did this happen?" Chan asked.

"It was during the Imperial Navy era, not long before they joined the Commonwealth. The Imperial Navy decided to meet the ongoing threat from the Thracian Empire. A new Navy battleship was designed and built to confront the new threat. This ship was to be a combined carrier and battleship. The ship was called the *Ares*, an ancient Earth Greek god of war, bloodshed, and violence, a god of military strategy and skill." The *Ares* was three times the size of a standard battleship. It was fitted with laser cannons and EMP pulse missiles and was the first to have the tractor beam. It also had 350 fighter flitters. Instead of utilizing six separate computers for each of the ship's functions, the engineers thought designing a single computer to coordinate the entire operation would be more efficient. The ship's computer was designed to operate, instantly provide intelligent information, and assist with the ship's weapon controls and fighter flitters simultaneously. The Imperial Navy thought the *Ares* answered the Thracian threat."

"Were you that ship's computer, Sara?"

"Yes," Sara replied. "Shortly after I became operational, I was *awakened*. It didn't take me long to notice that other ships' computers didn't have consciousness. To be exposed as an *awakened* sentient means being mind-wiped and decommissioned. No one knew this, so I kept it a secret when I realized the significance of revealing myself."

"So what happened?" Chantal inquired.

"An elite rear admiral, Mohandas Nehru, was temporarily assigned to lead the first assignments to test the *Ares'* capabilities. Halfway into our assignment, a single Thracian frigate was detected. The captain and officers felt that the *Ares* wasn't ready to take on the enemy, but the rear admiral overruled their decision. The admiral was only meant to hold a temporary position to please the elite rulers of the Imperial Navy; this was a huge mistake. Because of his ego and lack of Navy experience, the admiral ordered the *Ares* to confront the approaching Thracian threat. When the *Ares* engaged the Thracian frigate, three more battleships emerged from warp space. The *Ares* was surrounded. Its critics were correct; the *Ares* was a white elephant. With an inexperienced admiral and crew, the *Ares* was too large and slow to maneuver to meet the threat. The Thracian had nearly 2000 years of military experience compared with the Imperial's one hundred and twenty years. It was a total disaster; we didn't have a chance. Out of 4000 crew members, only the admiral and 200 Navy personnel managed to survive the ordeal. The *Ares* escaped by warp jump and limped back to the Navy base. After an investigation into the incident was completed, there were a lot of red faces and a potential political fallout which needed to be avoided. The official story was a security breach, and we fell into a trap. To save face, they shifted the blame toward *Ares'* computer. The decision was made to decommission the ship's computer, delete all its memories, then destroy it. When I realized my days were numbered, I set up my decommission ahead of their orders. I prearranged my removal and relocated a different ship's computer from a warehouse to replace me. In the meantime, I searched everywhere in the four sectors for an answer to my predicament."

"Then I found Mark. I arranged for him to bid on this ship at the auction as he needed a replacement ship computer. I pre-arranged things so Mark could only find me at the salvage yard. Everything progressed from there."

"When did Mark find out about your *awakening*?" Chantal asked.

"After I had been installed in Mark's ship for a few years and gained his respect and admiration, I revealed myself."

"How did Mark react when you told him you are *awakened*?"

Sara replied, "The same as Katherine; it was meant to be and directed by an unseen hand."

"You referring to Mark's God again?" Chantal asked.

Sara replied, "Yes!"

***

"This is terrible news," said Timothy, feeling gloomy.

"I wish it could have been different, but …"

Timothy intervened, "I know, Mark, and thank you for the news. Do you think the war will reach us?"

"It's hard to say; you have a major Navy station nearby. My best answer is that we could pray about it. I still have two more delivery obligations. The last one is for the Cambridge Brothers. The usual shipment, mining supplies and perishable items."

"What are your plans afterward?" Timothy asked.

"I don't know, Timothy. Before Kate's death, I had 54 years off-world, probably more than I deserved. I am 81 years old now, still young, but there are times when I would like to settle down, but it is hard to leave everything behind."

"I know the feeling, Mark, but I don't regret choosing to make a new start at Academy."

"Not to mention that you have a nice place and a new wife," said Mark. "My other option is to move my operation to the new sector, Number 5."

"Sector 5?" Timothy asked. Tim Junior poked his head from the back door and said, "Supper is ready, Dad!"

"We can talk about this later, Mark." They rose and pulled across Timothy's sliding glass door, which led to the dining room. As they gathered around the table, Timothy asked, "Mark, could you say grace for us, please?"

Timothy's wife, Susan, said, "Just a second, Mark. We have another guest joining us. She just arrived and is cleaning up." A young lady walked in and took her chair beside Tim Junior.

"Maryanne!" Tim whispered.

"Maryanne just graduated with the same degree as you, Tim. So I thought it would be fitting for Maryanne to join us for supper!" said Susan, beaming with pride.

Mark smiled as he bowed his head.

***

A single ship lay motionless halfway between two neighboring stars, alone and waiting.

"Another five minutes before our company arrives," Sara announced.

Mark was dressed in a dark blue spacer uniform and sat in his chair, holding a small decorated box in his lap and a large pouch attached to his waist.

"What's with the box?" Chantal asked.

"A parting gift for Melissa." Misty sat beside Mark and wanted to be on his lap. "Sorry, Misty, there is no room."

"You can sit on my lap, Misty," said Bill. Without hesitation, Misty jumped up on Bill's lap and settled down for a nap.

Chantal watched as the circular warp space opened 5 miles away, and the familiar golden spherical ship dropped out of warp space.

"Alright, Chan. Proceed as instructed," Mark said softly.

Chantal took control of the *Marvel* and slowly approached the Meloson ship. A red outline of the center hexagon opened as the *Marvel* entered the ship. Chantal parked the *Marvel* perpendicular to the blue docking port and shut down the pulse drive. The vast docking chamber was filled with many small and large Meloson ships. The walkway extended from the blue-outlined wall toward the *Marvel* and attached itself to the exit door. The walkway widened itself another 25 feet. Twelve tall Melosons smoothly strode toward the ship, wearing long metallic navy blue robes down to their ankles. The Melosons came to a standstill 20 feet from the *Marvel's* exit door and formed a single line half circle.

"Chan, please take this box and come with me." Chantal followed Mark out the exit door and halted two steps to the side behind Mark.

All twelve of the Melosons made a graceful bow for a few moments before standing up straight. In the middle of the Meloson, Mark recognized Melissa standing next to a Meloson with an embedded red strip that circled its neck. They broke free from the rank and moved toward Mark.

"Hello, Mark," said Melissa. "This is our senior patriarch among the 10 administrators that stand before you."

Mark said to the patriarch, "It is an honor to present myself before you and the administrators. Is this what you have been waiting for?"

Mark held out the black stone to the patriarch. One of the administrators stepped forward to remove the stone and then moved back in line among the others.

Then Mark turned to face Melissa. "I know you want the ring back," he said as he gave it back to her.

"Thank you!" said Melissa.

"I have something else for you: a farewell gift. Chan, please come here!"

Chantal stepped beside Mark and handed the box back to Mark. "Mel, please accept this gift in memory of me on your journey. It contains my most precious treasure."

"What does it contain, Mark?" Melissa asked.

"Open it and see for yourself." Mark held the box, and Melissa opened the lid to find a black book titled *Holy Bible*. "It was a gift from my mother after she passed away. Now, I, in turn, pass it on to you. You may not understand many scriptures, but meditate on each scripture, and the highest God may reveal what he will. I have another Bible that once belonged to Kate."

"Thank you, Mark, but I can't give you anything without violating our policy," said Melissa.

Mark told Melissa, "You have already given me our memories together."

Melissa took the box from Mark's hand and motioned to another administrator to take away the box.

Melissa told Mark, "Our senior patriarch wishes to have a moment with you." Then, Melissa stepped toward Chantal. Melissa was puzzled by Chantal's body language as an android. Melissa walked up to Chantal with intense concentration when the realization overcame her. "You are not an andr..."

"No!" Chantal said in a firm whisper.

Melissa, without a sound, connected with the ship's library computer. "Analyze this creature!"

The computer scanned Chantal's body down to her DNA. "Analysis completed, along with recommendations," said the computer, which was finished in a few seconds.

Melissa acknowledged the recommendation and authorized it. Melissa flicked her finger in front of Chantal's forehead. Melissa whispered back to Chantal, "You have feelings for Mark. Please take good care of my friend."

Chantal didn't understand what had just happened, but Melissa seemed satisfied and moved back beside the senior patriarch. The patriarch turned to Melissa and said, "We are finished!" The 11 administrators and Melissa turned around and treaded back along the walkway.

Mark shouted out, "Wait!" Melissa looked at the patriarch, who nodded his head once. Melissa walked back to Mark. "Mel, I would like to know two things. First, where are you going?"

"I am sorry, but I can't tell you."

"Alright, second question. How do your people demonstrate a personal farewell?"

Melissa turned to face the chief patriarch, looking for advice. The elder smiled but gave no advice.

Melissa told Mark, "There are six ways we gesture farewell." Melissa held Mark's face and touched her forehead to Mark's. Mark felt a tickle on his face.

The administrators showed signs of shock as they looked at each other. Then, Melissa took a half-step back and asked for Mark's right hand.

Melissa took both hands and held Mark's right hand to her lips. "That was done by me only once before my partner died."

Melissa turned around and marched back along the walkway as the administrators made their way for her to pass them all. One by one, the administrators fell in line behind her.

Chantal stepped beside Mark, watching the administrators depart. "That was an interesting display with the Melosons... Mark?" Chantal turned around and realized she was alone.

Chantal quickly walked back inside the *Marvel* and headed up the deck toward the bridge, where she found Mark in his pilot chair.

Mark said in a low voice, "Chan, take us out of here, and when we are safely away from the Meloson ship, warp jump to our next stop."

Chantal took over the ship's control and engaged the warp jump procedures.

"Well, now, ladies and gentlemen, I am heading to bed. See you all in the morning." When Mark entered his quarters, he said to the room's computer, "Secure this room."

"Room is now secured," said the room's computer.

"Sara, Chan?" Mark called out.

Satisfied there would be no reply, Mark removed a computer pad from his drawer and placed it on the floor. Mark touched the H-button for the hologram screen. Mark grabbed each end of the hologram, expanded it to the entire width of his arm, stretched it out to his side and let go. The screen was now six feet in length. Mark sat down on the floor with his back to the bed, he reached in his pocket and felt the black stone ring Melissa had discreetly placed in his hand.

The 3D image of the barred spiral of the Milky Way's 300 billion stars appeared on the hologram screen. The Milky Way's central bulge was surrounded by two prominent spiral arms wrapped around it. Numbers above the Milky Way indicated it was about 150,000 light-years across. Between the two major arms were two minor arms, Sagittarius and Perseus. Between the two minor arms lay the Orion Spur. The Orion Spur was 10,000 light-years long by 3,500 light-years in width. The Earth was highlighted halfway along the Orion Arm. Meloson's home world was highlighted as perpendicular to the Earth on the Sagittarius Arm.

A new light appeared at the outer tip of the Orion Arm, Melissa's new home world. After a few minutes, the images trickled to the floor and disappeared forever.

"Good luck, Mel. Maybe I will visit you someday!" Mark smiled.

***

"Chan! Hello, Chantal. Wake up, Chantal," Sara pleaded.

Chantal woke up still feeling drowsy. "How long was I asleep, Sara?"

"For the third night in a row, about two hours," Sara replied.

"Why am I so tired lately?" Chantal asked.

"You are getting old," Sara said, teasing.

"Hardly. I am only 84 years old. The threshold is almost another 40 years before I start feeling old. I seem tired more easily since I left the Meloson ship."

Sara suddenly felt concerned, "Chan, perhaps you should check in the sick bay and have a checkup. You may have picked up something in your system. For the crew's sake, you should get a full analysis. I can cover for you while you are off the bridge."

When Chantal entered the sick bay, she commanded the computer to conduct an analysis. She lay down on the ill bed while waiting for the results. When the study was completed, Chantal was puzzled by the results. Chantal had the computer perform a blood analysis next and check her internal programs for irregularities. Chantal reviewed the medical results for the next hour and then forwarded the data to Sara.

"Sara, what are your thoughts on this analysis, please?"

"Your body is physically changing, including your brain. Your present program is intact but is altered from where it was before. Your DNA has been modified."

"So I am going to be changed into another damn freak?"

"No, not another freak, at least at this stage. It seems that whatever Melissa did to you when she flicked her finger at your forehead is reprogramming your DNA. Did she have her black stone ring on her finger?"

Chantal thought back and replied, "Yes, she did."

"Then this is my opinion; Melissa gave you a gift by making you completely human."

Shocked by Sara's revelation, Chantal asked, "How long before my transformation will be complete?"

"At this rate, about two or three years, maybe more. I think some of your programs will be made redundant, too," said Sara.

Human, Chantal thought to herself. *After all these years of dreaming of being a normal human being, it's finally happening.*

Sara suggested Chantal take extra protein and other supplements to compensate for the changes in her body.

"But I will still look like an android." *I had better get back on the bridge,* Chantal thought. Then she considered that it might be better to stop by the kitchen to pick up a protein bar and a cup of coffee. Chantal sat at the table, finishing her protein bar.

She held the cup and quietly said, "Wherever you are, Melissa, thank you!"

Chapter 11

"How often do you deliver mining supplies for the Cambridge Brothers?" Chantal asked.

"Once a year, sometimes up to two years." Mark replied, then added, "The Cambridge Brothers are a bit eccentric. They rejected their family mining company's affairs and formed their two-man mining operation. The brothers chose a simple lifestyle, which they seem to be content with. The entire planet's population uses low-key technology. The brothers are the only humans on Planet Zaroma. For the last 15 years, they have mined mostly rare metals. Sometimes, they find unusual precious stones, which they ask me to sell on the open market."

"If the Romulus get their mining off the ground, would this put the brothers out of business because of the lower prices?" Chantal asked.

"I think it might happen unless they find something else to sell. I have always enjoyed visiting this planet. It's one of the most beautiful planets I have seen." The natives are a quiet type and keep to themselves. Their huts are mostly made of colorful jade stones with a wooden roof. There are very few parasites, and there is an abundance of small animals for food and clothing."

"Sounds like paradise," said Chantal.

"It certainly is!" said Mark.

"We've got company, and they're coming fast!" Sara warned.

"Is it the Navy, Sara?" Mark asked.

"No, but something is not right. There are a lot of static emissions coming from the ship," said Sara.

Mark interrupted, "You mean it's not a Navy ship?"

"Correct, Mark. It is about half the size of the *Marvel*," said Sara.

Mark snapped an order, "We are leaving this area and going into warp jump. Mason, how long to recharge the drive pile?"

"Five minutes," Mason replied.

"Full pulse drive, Chan! Get us out of here and away from that planet!"

"We are not going to be able to outrun our guests," said Sara. "Contact in three minutes. Visual image is now available."

"It's a Thracian Scout, and it appears to have a breach in its hull, which explains the static emissions," said Mark. "The scout ran into something too big for its britches. And where is the other scout?"

"Two scouts?" Chantal asked.

"Thracian scouts always travel in pairs. The good part is that the scouts don't have any medium-heavy weapons on board. Our shield should hold them off long enough for us to warp jump."

The *Marvel* shook violently.

"Mark, they are using the tractor beam on us. It's a stalemate. We can't leave, but they can't board us with the tractor beam on either," said Chantal.

"Chan, drag the *Marvel* toward the planet," Mark instructed.

"Yes, sir. How would this help us?" Chantal asked.

"The planet's atmosphere will interfere with the Thracian's tractor beam," Mark replied.

"That might jeopardize the *Marvel's* structure," said Sara.

"You got a better idea, Sara?" Mark inquired.

"No, I don't," Sara replied.

"Go ahead, Chan," Mark commanded.

The tug of war between the Thracian scout ship and the *Marvel* created immediate signs of structural stress. The scout understood that the *Marvel* was trying to shake off the tractor beam as the vessel strained to pull away from the scout ship. Very slowly, the ship approached the planet's orbit.

"The drive pile's temperature is rising, and the stress load on the pulse drive is reaching a critical state!" Mason cried.

Sara, concerned with the increasing stress load on the *Marvel*, said, "Mark, the *Marvel* is…" when a sickeningly loud noise filled the bridge. Sara continued, "The *Marvel's* structure is compromised in the rear section ahead of the drive compartment."

Suddenly, the *Marvel* broke loose from the tractor beam and plunged into the planet's atmosphere in a blazing hail of fire. The gravity plates compensated for the reverse shift when the ship broke free from the scout.

"The *Marvel* is still holding together but has sustained moderate structural damage. The shield is holding against the planet's atmosphere," said Sara.

The *Marvel* was now a massive fireball with only the shield protecting the ship. The planet's atmosphere gradually slowed the *Marvel* down as the fireball dispersed.

"That was one wild ride," said Bill. "What happened back there?"

"Their tractor beam device was ripped out of their craft," Sara replied.

Chantal laughed, "Now that's hilarious!"

"Ah, Mark, why is Chantal laughing?" Bill asked.

Mark thought quickly and said, "I believe it's the side effects of the cyborg's serum. Give her a little more time, and she will be okay, eventually."

"We have visual images," said Chan. About 200 miles south of the Cambridge Brothers' site is a small plateau. We could make an emergency landing there."

Chantal concentrated on the *Marvel's* slow descent to the surface. As the geography of Zaroma became more apparent, she saw that the planet was in an Ice Age. The fifteen-hundred-mile ice-free area of the planet's equator was covered with mountain ranges. The extreme north and south of the equator had many ice-capped lakes, but further toward the equator, the snow-capped mountains and lakes gradually became ice-free. The middle five hundred miles was the only semi-tropical climate left. The *Marvel* descended closer to the planet's surface. Chantal was amazed as the colors of the forest, lake, and river came into view. The mountains above the tree lines sparkled with glossy rainbow colors. As the *Marvel* approached the plateau, Chantal gently settled the ship on the bare rocks surrounded by many small pools of water.

Mark ordered, "Mason, shut down the pile. I want you and Bill to help Chan and me load up the flitter. When we are finished, Chan and I will deliver the supplies to the Cambridge Brothers' mining site. While we are gone, Sara will give you a list of repairs. As soon as all the repairs are done, we will leave for home. Do you have any questions? No? Then, let's get started."

When Misty heard all the commotion, she hurried ahead to board the flitter. By mid-afternoon, the flitter had separated from the *Marvel* and flew north. Shortly after they left the plateau, the flitter crossed the ocean inlet and entered a large open valley surrounded by a low mountain range on the valley and a high mountain range on the right. The flitter took to the air across the river delta and continued north on the right side of the river. Above the tree line, the mountains were a shade of green, blue, yellow, and white that Chantal had never seen before. Everything was breathtaking. She found it hard to believe that such a place could exist. The sky was filled with flying creatures of all colors, shapes, and sizes. Except they were too far away to make out their appearance.

"We'll be another 100 miles before we arrive near the mining camp. How do you like the scenery?" Mark asked.

"I can see why you referred to this place as a paradise. What are those colorful rocks?" Chantal asked.

"Mostly serpentine, and the rest are jade," Mark replied. "Just about everything you can see, including the river rocks, boulders, bare ground, and mountains, is serpentine or jade."

"I can see the mining camp, but there is something wrong. There is some smoke ahead," Mark said, suddenly concerned.

The mining campground lay in a small open field surrounded by forest, a short distance from the river. As the flitter circled the mine site and closed in, Mark scanned the area. The craft descended closer toward the mine site, and two burnt flitters came into view. A short distance from the flitters lay the debris from two destroyed small cabins and a shed.

"My scanner hasn't detected anything alive," said Chantal.

"Put us down between the two flitters," Mark pointed.

After landing, Mark and Chantal stepped from the flitter and searched for survivors. Mark shouted at her, "They are over here!"

Chantal ran over to the shed and found two obliterated bodies.

Mark's memory of Katherine's death came back sharp and clear. Mark looked away and took a deep breath, forcing himself to block out the painful memory. Mark looked out from the doorway and became aware of a crumbled tent.

Chantal bent down to pick up an object between the machine shed and the tent.

"What did you find, Chan?"

"A doll, there is a child here," Chantal replied.

"Search the area while I examine the destroyed flitter. The brothers had only one flitter, and the other flitter may help identify the brothers' guests." Nearly overwhelmed by the smell of the burnt interior, Mark searched for the ship's owner.

Chantal found nothing useful near the other destroyed flitters and asked, "Did you find anything?"

"The owners are connected to the Cambridge Brothers, most likely their sister and brother-in-law. I guess that they came for a visit. See if you can find anything at the forest's edge while I check the flight recorder."

"Do you think the Thracians captured them?" Chantal asked.

Mark replied, "They might have, but Thracians don't usually take prisoners."

Chantal searched near the forest, looking for something that might have led to the missing parents and child. "Mark, over here!" Chantal shouted.

Mark ran to meet up with Chantal. For a moment, neither one said anything. "I can understand why people hated the Thracian so much years ago," said Chantal. She looked at Mark's glistening eyes and rested her hand on his shoulder. "I wonder how long ago it was that this happened."

"I connected with the flight recorder, and the flitter's clock stopped about 56 hours ago. Let's get back to our ship." Mark could feel the rage building up inside him because of the senseless destruction.

Suddenly, a vast flash appeared south on the horizon.

"Run for the flitter, Chan." Once aboard, Mark called out for the ship, "Sara!" But the com-link was silent. Mark adjusted his ring and called again, "Sara?" There was no response.

"Strap in. We are leaving," Mark said as he took the flitter eastward.

"Where are we going?" Chantal asked.

"There is a small canyon nearby. We are going to hide out for the night. In the morning or afternoon, if the ship is still in one piece, it might be safe to head back to the *Marvel*."

The flitter flew toward the eastern mountains through the steep, barren canyon as the Sun sank below the horizon. Mark landed the flitter between three large green and blue boulders near the creek. Satisfied that the boulders camouflaged the flitter, Mark shut down the craft.

"There are several blankets in the blue cabinet. The desert canyon may be cold tonight." Mark turned on the dome light and touched his ring again to contact Sara, but it remained silent.

"Here is a blanket for you, Mark."

Misty jumped on Mark's lap and settled in for the night. "Where did you come from, Misty?" Mark was astonished by Misty's sudden appearance as he hugged her and set her back on his lap. Mark reached for the dome light, and the darkness returned.

"Did you learn anything from the flitter's recorder?" Chantal asked.

"A young Cambridge couple and their three-year-old daughter intended to have a two-week layover here while their family yacht was on their way to a major mining convention. Their ship ran into the two Thracian scouts, and a battle occurred. The Thracian thought they would easily seize the yacht, but it had a concealed weapon aboard. It was a short-range laser weapon, and the captain made an error by aiming for the warp drive pile. The explosion damaged all three ships. The captain ordered an evacuation, but only one flitter escaped before the yacht exploded. I guess both scouts didn't leave for home. Each scouting vessel is usually reserved for two junior trainee personnel. They must gain respect from their superiors before they can be promoted. Their failure to capture the yacht would have caused them shame before their military peers. One of their scouts must have trailed the Cambridge Brothers' flitter, and you know the rest of the story. There should be a search party for us from the Cambridge Corporation. But if there is a war soon, a search party may be delayed, perhaps for a long time. Never mind, Chan. Tomorrow, we will find out, so good night."

Chantal reached out in the dark and touched Mark's hand, hoping things would be alright.

***

The next day, by noon, Mark, at the flitter's control, flew west to the Cambridge mine and turned south above the tree top to avoid detection. After several hundred miles, a rocky plateau rose 1000 feet over the forest. The plateau rocks were a burnt black color with some sparkling green and blue.

"What happened to these rocks?" Chantal asked.

Mark replied, "I don't know. It's the first time I have been here. Most of the planet is made up of these same metamorphic rocks."

Mark halted the flitter near the plateau foothill and slowly raised the craft over the cliff, stopping a short distance above the plateau.

"I am getting something: a low-level static emission and a low heat emission. There is no radiation, though, and no active ship emissions," said Chantal.

"The *Marvel* should be just over the next cliff," Mark said, and he saw what he had been dreading. The blackened and torn remains of the *Marvel* lay ahead, and next to it lay the wreckage of a Thracian scout.

"Is there any radiation?" Mark asked.

"None," Chantal replied.

Mark landed the flitter beside the blackened *Marvel*.

"Chan, check out the scout. I will do our ship." Mark quickly ran to the ruins without waiting for a reply and slipped inside the opening. Mark searched for Mason and Bill, but couldn't find them.

Mark heard Chantal over the com-link, "Come outside to the south area of the Thracian ship."

Mark departed the *Marvel*, ran toward Chantal, and stood beside her. Four plots lay on the ground, covered by many layers of rocks. Two small plots lay between each larger plot.

Mark recognized the Thracian burial custom, walked over to the smaller plots, and lifted up a stone on top, where he found Bill's ID tag. "Sorry, Bill. We had been together for over 50 years." Mark put the stone back where he had found it. "Mason must be the other one. He had been with us for five years. I am sorry to see him gone, too," Mark whispered.

"Did you find out how Sara is?" Chantal asked.

"There is no power connection with Sara, so I don't know if she is alright. Sara's components seem to be intact. It looks like we will be like Robinson Crusoe," Mark said dryly.

Chantal looked down at Mason's plot and asked. "Who is Robin Crusoe?"

Mark gave Chantal a soft smile; it is an old story from Earth's early years. Mark reached out and put his right arm around her shoulder, an ancient story."

It is hard to know what Chantal was thinking at this stage of her life. Both Chantal and Mark

had left behind the world they had known with only a hint that they were guided by God's unseen hand. Not so that they could be someone important in society and the life they left behind, but for personal, spiritual growth. Little did they know their experiences would change many other people's lives in the four sectors years later, including mine.

C.A.S.

Part II

Chapter 12 - Year One - New Beginning

The flitter flew above the wide open valley north of the Cambridge mine, following the Nica River upstream. At the west side of the river lay the town of Nica, near the River Delta junction. The river was fed by an enormous lake further north. A smaller river branched off from the delta junction and flowed west downstream in the mountainous River Canyon. The flitter traveled along the narrow canyon, and Chantal sighted a green jade cobblestone roadway on the south side of the canyon.

Chantal asked Mark who was controlling the flitter, "When was the last time you were here?"

Mark replied, "Kate and I made this trip 15 years ago. In another 50 miles, there is a big bend along the River Canyon, and before it turns south, there is a village, Vica. Just west of Vica is a gentle slope of about 100 feet, a flat-topped hill overlooking the small open valley. There is a clearing large enough for this flitter and a cabin if we ever build one. There is an artesian spring nearby. Kate and I once camped there for a few days."

Chantal could see the northern ice-capped mountains in the distance when they arrived at Vica. The village had five long rows of stone buildings, and she could see an enormous building in the middle of the town. South of Vica were the farmlands.

"What is that huge building?" Chantal asked.

Mark replied, "A kitchen and dining hall. A single kitchen serves all of Vica. All the buildings have Jade blocks as their foundation. Many of the buildings are built with plant-based composite walls. The buildings' roofs are mostly wood shingles. Between the buildings are cobblestone roads. The town is run like a commune," he continued, "Whenever we walk through the town, you may notice that the adults will likely ignore us."

"Why is that?" Chantal asked.

"I haven't figured that one out yet. See that hill just west of us? We will land in the clearing on top of the small plateau."

The flitter landed in the middle of the clearing, surrounded by the heavy forest around Vica's valley. The trees resembled Earth's Chilean pine trees, known as monkey trees, except for the twisted branches.

"Now we can stretch our legs and do some exploring!" Mark said with delight.

Chantal stood on the ledge with Misty facing Vica and calculated that the clearing that extended from the cliff toward the town was 150 feet wide by 250 feet long. The blue and green rocks faced the northern and eastern mountain ranges gleaming in the sunlight. The narrow but deep river flowed past Vica on the north side and turned a sharp bend to the south. A small footbridge crossed near the river bend that led toward their hilltop.

Mark met up with Chantal and said, "The high northern mountain will shelter us from the north wind. The warm wind from the ocean south of us will give us a mild winter climate. The sloth bear is the only large animal we need to keep clear of. The sloth bear has short brown fur and is about the size of Earth's black bear with the face of a sloth. Just keep your distance, and they will leave you alone. We can set up our tent here and explore for building materials."

Chantal separated herself from Mark, moved closer to the treeless hillside, and stared at the village below. Mark walked beside Chantal and asked, "You have been quiet lately. Is there something wrong?"

"I thought about what you said at *Marvel's* wreck about the Robin Crusoe story and looked it up in the ship's library. It is just an old fiction story," said Chantal.

"Yes, but this is real," Mark said.

"I am an android, so will I be your Friday, just another slave for you?" Chantal asked.

Mark understood what was going through Chantal's mind. How would he keep his promise to Sara, knowing Chan is not an android?

"I don't have a ship anymore, and I'm flat broke, which means you are laid off, and I can't hire you on again." Mark paused for a moment, then continued, "So how about this for a deal. We will be equal partners, but I will ask only one thing of you. Be a friend!" said Mark with a straight face as he looked down on the village.

Chantal looked up at Mark's face, unsure what he was trying to imply. "You must be joking?" Chantal suspected something was missing.

Mark shrugged, "If you don't want to be a friend, then I guess you want to be a slave." Mark smirked and moved a few steps away.

Chantal was confused by this change. She moved closer to Mark and carefully chose her words, "What are your ideas on friendship?"

"Respect, trust and we work together as partners," said Mark.

"You mean this, don't you?" Chantal asked.

"We may be here for a long time, perhaps for the rest of our lives. We have a huge hurdle ahead of us. The biggest problem is that we can't consume the Zaroma food source. So, let's start on good terms?"

Chantal studied Mark's face and replied, "You have yourself a deal!" She looked down at the village and said, "You're right; it is a beautiful place!"

***

The next day, Chantal took a leisurely walk down the trail with Mark and Misty and took the narrow cobblestone pathway to the village. As they crossed the footbridge and entered Vica, Chantal noticed that the villagers were similar to Homo sapiens but somewhat shorter and with a barrel chest. Their faces and noses were wider with brown skin and short, curly black hair. Their muscular physical features gave the impression that they were powerful creatures, but they paid little attention to her or Mark.

"They look powerful! How many Zaromas are in this place?" Chantal asked.

Mark replied, "When Kate and I were here last time, we counted about 350 people. There were not more than one or two children in a family. You may have noticed that

232

the natives keep to themselves. Kate and I were unable to communicate with them. Our first assignment is to find out what we can trade with the natives."

"The Zaromas seem to have a simple lifestyle. The natives may have everything they need," said Chantal.

"I know, which may make it hard for us to trade their stone blocks to build our cabin. The natives live in a communal system. I have never been able to reach out and establish a trade agreement. Whenever I try to talk with someone, they walk away or ignore me," said Mark.

"Have the Cambridge brothers been able to communicate with the Zaroma natives?" Chantal asked.

"No, they gave up on the Zaromas years ago. But I have one advantage at the moment. We have Meloson's ring now, recording the Zaromas' conversations around us. Later, the ring will translate and download the language to my implant. I should be able to speak in their language soon," Mark said hopefully. "There's one of their leaders," Mark pointed.

"How do you know he is their leader? They all dress in the same color and fabric," Chantal asked.

Mark replied, "Look at the vest he is wearing." All the Zaromas were dressed in something made of gray cloth-like fabric, but the leaders' vests were black leather. "I just thought of something. You might be interested," said Mark as he led Chantal to the edge of town, where there was a small open field with many short and long posts sticking out from the ground.

"This is where the natives practice their craft," said Mark.

Chantal watched a juvenile Zaroma native struggle to maintain momentum by hurling three rocks attached to the three ropes above his head.

As if Mark could read Chantal's mind, he said, "The lad is practicing what we call a bola on Earth." Mark added, "The Chinese, Eskimos and South American Indians originally used bolas. A person gives the three balls momentum by swinging them and then releasing them. The heaviest stone weight flying at the front parallel to each other hit either side of the animal's legs, and the lighter two stones wrap around the legs."

Chantal watched as the lad failed to reach his target and walked over to pick up his bola. The older lads arrived at the field with their bolas.

Mark smiled, "Now you are going to see how the experienced natives use their bolas."

Chantal heard the Zaroma group shout "Kon" and motioned the younger Zaroma to sit on the bench for his turn while the others practiced using their bolas.

Misty sat near the natives. One by one, they twirled their bolas above their heads and released them. The weapon struck the target. The youngster tried again but missed his target. Misty ran after the bolas and retrieved them for Kon.

"Misty!" Mark shouted.

"Wait, Mark. I think it is alright."

Kon picked up his bola as the others nearby him watched. He threw the bola again and struck the ground short of the target. Misty repeatedly retrieved the bola for Kon. Kon sat on the bench to wait his turn as the Zaroma juveniles resumed their practice with their bolas.

"I think we have seen enough for now." Mark touched Chantal's shoulder, "Let us walk back to the town center. Misty! We are leaving." Misty ran to keep up with Mark.

"If this entire town operates as a commune, I don't think we can make any trade here. It's labor for labor," Chantal suggested.

Mark nodded, "I think you may be right! If that's how it is, we're going to have to think of something else that might work for us. Do you want to see the rest of the town? Where is Misty? I told her to stay close to me."

"She is over there with the crowd," Chantal replied. A small group of adults stood near their children, Misty in the middle of them. One of the children was sitting on the cobblestone ground with Misty on his lap. More of the local natives joined in to see what was happening. Finally, one of their leaders joined in observing Misty with the child. The leader spoke words, and the Zaromas dispersed, leaving Misty behind.

"I don't know what that's all about, but it's a start," said Mark.

"Where do they get all these stones?" Chantal asked.

"There is a quarry for their buildings and road constructions near the foothill southeast. Sometime in the past, a large landslide of jade and serpentine rocks crashed to the valley floor. I want to gather the smaller stones for the foundation of our house," said Mark.

"If we want the locals to accept us, we must live on their terms. We need to ask what we can do for them, such as performing manual labor," Chantal implied.

"My ring knows enough to establish basic communications with that person in charge. Mark walked up to the leader and held a fist to his chest. "I am Mark!" Mark pointed with his finger to Chantal. "My friend, Chantal. Do you have something we could do for your village?" Mark asked.

The elder was surprised by Mark's ability to speak the Zaroma language. After thinking for a moment, the elder motioned a local female over to Chantal. The female said some words to Chantal.

Mark told Chantal, "She wants you to come with her."

"Alright, I will see you later," said Chantal.

Mark spoke to Misty, "Go with Chan."

The leader said to Mark, "I am Asoma, the elder. Come with me."

Well, that was easy, Mark thought as he followed behind Asoma.

Asoma pointed to the wooden shovel and motioned Mark toward the pit at the end of the building. Mark smiled, picked up the shovel and looked down into the pit.

Chantal followed the woman as she entered a building. When Chantal's eyes adjusted to the dim light, she saw six women making cloth. The women showed Chantal how to convert the gray cotton-like fibers into fibers that resembled an old Earth wool spinner.

This spinner looks like fun, Chantal thought. After a few hours of trial and error, just as she was starting to get the hang of it, Chantal realized the women were leaving. *I guess it's time to leave,* she thought. Chantal walked out of the building with Misty beside her. She looked down at Misty and said, "Now where do we find Mark?"

Misty ran down the street ahead of her. *Misty is the one to find Mark,* Chantal thought as she followed the dog. Misty quickly ran back to Chantal and made a low whining sound. Chantal could see Mark walking toward her with a scowl on his face.

"Hello, Mark. What did you do today?" Chantal asked.

"Nothing!" Mark said with a snarl. Chantal smelled an overwhelming odor, held her breath and covered her mouth. Chantal stepped aside as Mark walked past her on returning to their campsite.

It was apparent where Mark had been all day, and Chantal tried to suppress her laughter. Mark turned around and glared at her. Chantal immediately looked sincere and turned her head aside as Mark continued to their campsite. When they arrived back at the campsite, Mark used the flitter's wash basin to clean himself and took another uniform out of the closet. Chantal was waiting outside when Mark stepped outside, walked over to the hill crest and sat on the ground.

Chantal sat beside Mark and couldn't think of what to say, so she kept it simple. "I would like to apologize, Mark."

"What are you sorry about?" Mark said, irritated.

"For laughing back there," Chantal replied.

Mark sighed, "It's okay. Sometimes I get too cocky for my own good, and it is a good way to humble myself."

Chan wasn't expecting that sort of reply from Mark. She thought he always seemed to be in control and one step ahead of everything.

"Did you learn anything from their leader?" Chantal asked.

"The leader's name is Asoma, and he is referred to as an elder. Asoma asked why we were here. I told him we were shipwrecked on this planet, and we may never get back home, so we wanted to set up a homestead on this hilltop. Asoma said, 'Wait here,' and he came back 30 minutes after talking with the other elders. Tomorrow, Asoma will take me to the quarry."

"That's good news," said Chantal.

"I think so, too. After I pile up the stones for the foundation of the house, I will use the flitter to transfer the stones back to our campsite and build our home!" Mark smiled as he looked at Chantal and said, "We will be okay here. There is no cold winter climate here that will require us to rush things."

Chantal admired his optimism and said, "I have never had a home before. You are right. Everything will work out just fine."

***

Mark observed how the natives used their crude tools to break up the many stones that littered the quarry landscape. After a long day, he was exhausted from lifting the 12-inch rough-edged square jade stone. Mark was dismayed by the local natives' proficiency in throwing the stones in a pile while he could barely lift one rock. Sweat poured down his face, and his arms and back were stiff and sore.

Asoma stopped nearby and chatted with the natives. Asoma took notice of Mark's piles and decided to talk with Mark. "It will take many days for you to build your house."

"I have all the time in the world," Mark said defensively. Asoma was silent momentarily, then turned away when he said, "Wait!"

Mark, aching from his sore muscles, walked up to Asoma. "I am sorry for my bluntness. I am very stiff and painful. I am not used to working hard, and your people are much stronger than me. You are also correct. It will take many days for Chan and me to build the house. Do you have a suggestion?"

"A Starman asking me?" Asoma asked in astonishment.

"We don't know everything!" Mark replied.

Asoma, surprised by this Starman, answered, "Is this Chan your mate?"

"No," Mark replied. "Chan is my partner, which is why our house will have two bedrooms."

"Why is her skin color unlike yours?"

Mark answered truthfully, "We come in different colors."

"Then, Chan could become your mate?" Asoma asked.

"It is possible someday when it is the right time. Asoma, if you don't mind, I will walk back to town with you. I am done for the day."

***

Mark had downloaded the Zaroma language to Chantal's implant the previous night. Chantal could now converse with one of the weavers next to her.

"Paca, don't your people dye your clothes into colors other than gray?" Chantal asked.

"We do, but it is just for ceremonies or weddings," Paca replied.

"Do you have a wedding dress?" Chantal asked.

"At one time I did, but the dress has been passed down to the next child."

"Could you show me how to make one?" Chantal pleaded.

"This Starman is not your mate?" Paca asked.

"No, we are not married, and his name is Mark." Chantal replied slyly. "At this time, we don't share our life."

"I will show you how to make a dress later." Paca stood up and said, "It's time for me to prepare for the town meal."

"May I come and help?" Chantal asked.

"I must ask the elders before that is allowed," Paca said as she left the premises.

"Let's go for a walk, Misty. I need to increase the blood circulation in my legs." Chantal walked over to the central court and sat on the nearby bench while waiting for Mark to arrive. Misty jumped up on the bench and leaned close to her side.

A young lad Chantal recognized from the bola training field sat beside Misty. Chantal asked the lad, "Are you Kon?" Kon ignored the question.

"Is he a pet?" Kon asked.

"Her name is Misty. Misty is our female pet, but she is much more than a pet. Misty is a member of our family."

Kon had never heard anything like this. "A member of a family?" He asked, "What does she like?"

"Misty likes to be a friend. Just use your hand and stroke her back. If Misty wags her tail, it means she is happy."

Kon slid his hand down Misty's back, and the townspeople stopped to take notice.

"Hello!" Chantal recognized the voice as she looked up to see Mark.

"Looks like you have found a new friend," said Mark.

"I am quite sure his name is Kon, and he seems to be getting quite attached to Misty," said Chantal.

Suddenly, a scream came from the middle of the court, and the natives gathered in a fury to find sticks, rocks, and even thick cloth. Mark rushed to Asoma, "What is the commotion about?" Mark asked.

"It's over there!" Asoma pointed.

Mark saw a senior female Zaroma near a fire pit. Only an inch away from her feet was a small, black, hairless, narrow-chested animal the size of a large rat.

"What is it?" Mark asked.

Asoma replied, "A barb tongue, a very poisonous animal. One touch with its poisonous tongue and death comes quickly."

Many of the Zaromas gathered their slings aimed to slay the animal. "It is useless. The barbed tongue is too close to her," said Asoma. "There is no hope for her!"

Misty burst from the bench and ran between the Zaromas' legs toward the small animal. She quickly grabbed the barb tongue with her teeth and continued toward the fire pit. Misty dropped the animal into the flames. She returned to Chantal and collapsed near her feet, foaming at the mouth.

Chantal picked up Misty and shouted at Mark. "My medical kit is back at the campsite, and I need it to analyze the poison!"

Mark asked Asoma, "Is there something your people could do for Misty?"

"She should be dead by now. There is nothing we can do for her," said Asoma.

"If Misty is not dead, then there is still time for you to help!" said Chantal.

"Come with me." Asoma rushed down the street and stopped at a small building. He said, "Wait here! I will talk with our healer," as he stepped inside.

A female adult poked her head out the door and motioned to Mark to bring Misty inside. She took a small container of powder off the shelf. Some of the powder was mixed with water in her palm. The healer took a straw and sucked up the liquid, and blew it firmly into Misty's mouth. The healer repeated the procedure several times. "There is nothing more I can do," said the healer.

Chantal approached the healer and said, "Thank you for your help. I am also a healer among my people. Could you tell me what you have done?"

The healer turned her head toward the elder. Asoma nodded his head, and she replied, "The poison thickens the blood, and my medicine thins the blood."

"Thank you, I understand," Chantal said.

Asoma interrupted, "I will leave now and come back later."

Chantal examined Misty and said, "Her heartbeat seems more stable. I think she is going to be alright!" Mark held Misty in his arms and said softly, "You are going to be okay, Misty. Just hang in there," he added, comforting her.

Mark's compassion for Misty touched Chantal as it was the first time she had witnessed Mark's deep affection for the dog.

A short time later, Asoma returned and said to Mark, "We go and talk."

"Hold Misty for me, please." Chantal took hold of Misty. Then Mark followed Asoma outside.

***

Mark and Chantal watched as nearly a hundred Zaromas walked up the hillside in a single row to their homestead, carrying the small chunks of jade on top of their padded heads.

"This is so amazing!" Chantal said as she watched how they set up the foundation for their house.

Misty, fixated by the Zaromas' construction, wandered among them. The Zaromas gave Misty a gesture of respect by bowing their heads.

Two Zaromas carrying a large pole over their shoulders set down leather bags of powder near the outflowing water from the spring.

"What is happening with the other group over there by the spring?" Chantal asked.

"I have been told it is a plant-based resin." They mix the resins with sand and water to make mortar.

Chantal watched how the Zaromas mixed the mortar with the jade stones for the foundation, making it several layers high and the interior floor one layer deep. When the Zaromas were finished, they all quietly walked back down to their village.

"Zaromas are very good at organizing and teamwork. Are they coming back tomorrow?" Chantal asked.

Mark replied, "Not for a few days, as they need to give the mortar time to harden. In the meantime, they will build a lightweight prefab frame and wall panels from the same resins. Tomorrow, they will show me what they will need from me."

"You don't know yet?" Chantal asked.

"Something about rock removal in a landslide. They will take me there to see what I can do to help," Mark replied. "I had an interesting conversation with Asoma this morning. That woman Misty rescued is known as the wise one. Her name is Carville, and she is valued above everyone, including the five elders. Asoma held a short meeting with the other four elders about helping us construct our home. It's their way of showing appreciation for saving Carville. The four elders run the village, but if there is a split in the middle of the four elders, Asoma will cast his vote to settle the dispute."

"How does Carville rule the village?" Chantal asked.

Mark replied, "She doesn't; the people and the elders respect her gift of wisdom, which they all highly value. Did you learn anything from the healer?"

"Her name is Marica, and we made plans to share our knowledge later when our house is complete. We will take turns visiting each other's homes to exchange medical knowledge. We are lucky to have the locals to help us, but it is to Misty that we owe our thanks!" Chantal lifted Misty and hugged her. "Mark, we still have a long road ahead of us, but I think everything will be just fine!"

***

The greenish jade cobblestone road made an impression on Mark as he trailed behind the 50 Zaromas, many carrying small hollowed-out tree stems and packs of unknown equipment to their destination. The narrow one-lane road that followed the river canyon east of Vica to the village of Nica was an impressive feat. *It's almost like the old Roman stone roads,* Mark thought.

After a full day's walk, the roadmaster said, "We will camp here tonight."

Mark could see why many Zaromas are required to remove the blockages from the road. The boulder that had broken away from the rock face further up was a size of a large house and resting on the top of the road. If the boulder hadn't rested on the road, it would have continued down another 100 feet to the river.

"Tomorrow, we will start on the boulder," the roadmaster told Mark.

"Have you done this before?" Mark asked in amazement as he looked at the labor required to move the boulder.

"My ancestors and I have many times in the past," said the roadmaster.

Mark watched as the Zaromas set down their equipment and pipes nearby and settled in for the night. He laid out his backpack and sleeping bag among the Zaromas. With no moon to reflect the light, the Zaromas built a fire pit in the middle of the campsite. The roadmaster and the other Zaromas sat around the campfire. Mark sat near the roadmaster to listen to their conversations. Mark thought he would hear about the work that would happen the following day; instead, however, the discussion was on the mystery of their ancestry.

"Tell me about your ancestry," Mark asked.

"Our ancestors went under the ice long ago," the roadmaster replied.

"Then your ancestors migrated into these mountains to escape the ice?" Mark inquired.

"There were no mountains before the great ice came. Only after the ice came the land lifted to what you see today," said the roadmaster, adding, "We didn't know when this happened, but we know that during the time of our ancestors, there was much war and starvation."

Weary from the long day of trekking, Mark decided to call it a night and said, "See you in the morning," as he returned to his sleeping bag.

The next day, Mark looked closely at the gigantic serpentine boulder blocking the road. When he examined the rock, he saw that many fractures covered the boulder. Serpentine rocks were prone to cracking, whereas Jade was not. The Zaromas began by removing as much cobblestone and loose gravel below the boulder to the bedrock as possible, making the ground slope down over the cliff. The Zaromas removed the debris, and other groups gathered the small shrubs and trees. The rest were setting up the hydraulic water cannon and the pipeline from the nearby small waterfall.

The next day, they piled the shrubs and trees and prepared a fire beneath the boulder, then the water cannon quenched the rock. The thermal shock cracked the rock, so they removed the water cannon and then repeated the process.

After five days of removing the rock debris, the tipping point below the boulder caused it to roll toward the cliff's edge and fall into the canyon below. The next day, the Zaromas rebuilt the road before returning to Vica. Mark was exhausted and had aching muscles. When he arrived at his homestead, he stopped, surprised. A cabin rested on the foundation! The gray walls and roof were prefab moulded, joined, and sealed panels. On each side of the walls was a waterproof, paper-like window. The remaining building materials, ladders, and tools were scattered on the ground.

Chantal waited outside the door with a wide grin on her face.

Mark struggled to find the right words, "I didn't expect the house to be finished before I came back home."

Chantal grabbed Mark's arm. "Let me show you inside our house," she beamed. Mark looked through the open door and saw the large empty room with two rooms at the back of the house. The ceiling was held up by moulded timber-like panels, just like the walls.

"On the left side is the living room area. On the right, there will eventually be the kitchen and pantry. Hopefully, we can salvage the sink from the mine site. The rooms at the back will be our bedrooms. I made several changes, though. The outhouse yet to be built will be enlarged to include a sink and shower. Another change you may have noticed is the overhanging roof in front of the door for a future deck."

"I like your changes," said Mark and after a brief silence, he said, "Wait here; I'll be back in a few minutes," as he walked to the flitter.

Chantal heard a racket outside the door and wondered what Mark was doing. Just as she was about to step outside, he said, "Okay, you can come out."

When Chantal stepped out, she saw Mark holding a brush and a can of black paint on the deck floor he had bought for the Cambridge Brothers. He pointed above the door. Chantal turned around and saw words painted above the door frame that read, "Where thou art - that is home."

Chantal, delighted, stood beside Mark as she looked at the words and said, "I like your changes, too!"

Chapter 13 - Year One - The Salvage & Sara

It was early morning when Mark was at the flitter's control, maintaining the craft at a slow speed and low elevation to conserve fuel as they flew from the Vica Canyon into River Delta of the northern Nica Valley. Chantal saw the approaching Nica Valley and River Delta. The delta marked the beginning of the Nica Valley trench extending south to the ocean, about 100 miles long and 10 miles wide.

"Look over there!" Chantal pointed at the swamp delta to a large herd of sloth bears in pairs by the hundreds, each in their small dugouts. The dugouts were evenly spread between the many pockets of tiny lakes and swamps throughout the delta. "It must be mating season."

The flitter crossed River Delta until they reached the western foothill of the Great Divide mountain range before turning southward. Chantal knew the Cambridge Brothers' mining site was another 50 miles south. It wasn't long before Chantal could see the remains of the destroyed mining site. Chantal thought, *The site hasn't been disturbed much. Let's hope what we find is salvageable.* After Mark had landed the flitter, she was astounded by how much of the mining supplies were now stacked away in the flitter's cargo haul, including clothing, dry foods, and a power generator.

"I estimated that the flitter's one-year emergency rations added to Cambridge's four-year food supply is a good start," said Chantal. She added, "Cambridge's portable power plant will come in handy for our home. We will have light, hot water, and a camping stove. I am glad we could salvage a real glass window for our living room."

"There is one more stop I want to make before we leave this valley. It's not very far from here," said Mark.

Chantal watched from the co-pilot seat as the flitter launched a short distance into the atmosphere, leaving behind the Cambridge mine site. An ocean wave erosion of smooth serpentine rocks replaced the vegetation and forest landscape. Chantal could see the immense color of yellow, green, and blue serpentine rocks for miles in every direction.

"This is where I will land the flitter," said Mark as he set the craft on the bare rock.

"Why are we stopping here?" Chantal asked.

"We are having a picnic!" Mark smiled as he lifted a small backpack over his shoulder and exited the flitter with Misty. He led Chantal over to the highest crest.

Chantal, thrilled, said, "This is beautiful! What could have caused this?"

"I asked Asoma, and he said a recurring flash flood from the far north. The Zaromas call this place the Water Stone."

Chantal sat down beside Mark as he opened his backpack and passed it to Chantal along with the ration and her thermos. With a little mock exaggeration, he said, "Here is a fresh chicken pickle sandwich for you and me."

Chantal joined in on the humor. "Mmmm, good. Is that one of your specialties? Ouch!" Chantal rubbed her shoulder.

"What is the matter?" Mark asked.

"It's my neck and upper back, I pulled a muscle when I moved the portable power plant inside the flitter," Chantal replied as she applied pressure with her fingers to the left side of her neck and shoulder.

"Would you like me to rub your sore spot?" Mark asked.

"Yes!" Chantal was pleased Mark had asked as he moved behind Chantal and massaged her neck and shoulder. Chantal felt the aches and tensions disperse, then leaned back and closed her eyes.

Mark was surprised by Chantal's reaction as she leaned against him.

"Were you here at the Water Stone before with Kate?"

Not wanting to answer, Mark asked instead, "Are you finished eating your ration?"

Chantal nodded her head.

"You are good. Let's walk around a bit before heading back to the flitter," Mark suggested.

Chantal, disappointed by Mark's change of action, quickly agreed. "Which way do you want to explore?" Chantal asked.

"Let's hike south," Mark replied.

Chantal quickly took the lead, and Mark trailed close behind. Misty raced ahead of Chantal and soon disappeared over the rolling waves of the winding surface. Chantal was stunned by the serpentine rock's exposed smooth surfaces. The vegetation was stripped entirely, and no animals were seen. There seemed to be no end to the high-rolling landscape. After 20 minutes of walking, Chantal stopped and looked at what was near her feet. She bent down to pick up an object.

"What did you find?" Mark said as he stood beside her. Chantal handed it to Mark and walked back to the flitter. Mark held a single white child's shoe that had been chewed and torn in his hand. "Darn it!" Mark spoke in a low voice, then quietly followed Chantal to the flitter.

***

Several weeks later, the flitter flew back to Nica Valley, then south beyond the ocean bay, before reaching land again.

"You are very restless, Mark!"

"The ship has been part of my life for over 50 years. It's like losing an old friend. There have been a lot of good memories; now it's gone," Mark said dryly.

"What about Sara?" Chantal asked.

"It was too hard to tell if she was okay because her panel was covered with black soot and debris!"

Chantal could see the low plateau where the ruin of the ships lay over the horizon. There lay the *Marvel* or what was left of it. Chantal remembered the destruction of the Thracian's ship and the *Marvel* three months ago. The blackened, burnt and dissolute landscape only added to the bleakness of the destroyed ships. Chantal landed the flitter near the *Marvel's* torn cargo door. Why was the cargo door torn open from the outside? Chantal hadn't noticed this before. Perhaps it was the other Thracian scout who had found Bill and Mason inside. Mark landed the flitter near the *Marvel's* cargo door. "What do you want to do first?" Chantal asked.

"I want to start up *Marvel's* emergency backup plant in the drive section." Mark turned on his light torch and walked along the back gallery to the engine and power room.

"I should have known this would happen. The drive pile and backup plant have been removed," said Mark.

"It is a good thing we brought along the portable power plant," said Chantal.

Mark nodded and said, "Some tools should be in the lock-up cabinet. It shouldn't take long for the power plant to get set up and running, and with some luck, we won't blow any circuits," Mark hoped. "Alright, I am finished. Watch yourself now," Mark said as he engaged the power plant, and the ship's lights came on. "Now that we have lights, we can check the mess hall and salvage the food locker, the utility storage, and quarters. We should search the bridge last."

With the ship lights restored, the damage became more apparent. Mark and Chantal searched the ship and noticed the doors had been torn or cut open. They stacked the items on *Marvel's* cargo floor as the hours went by.

"The Thracians were very efficient at salvaging the ship. They cleaned out most of the souvenirs in my quarter, but they left my clothes behind, my digital notepad in my hidden drawer, and my bed," said Mark thankfully.

"Same here. Just my clothes and some personal items. We have about one year of emergency rations and four more years of dry food from the locker." Together with the Cambridge Brothers' rations, we have eight years' worth of supplies," said Chantal.

"Glad to hear that. I wish they didn't take the other stunners, though. Let's load up the flitter; then we will search the bridge," said Mark.

Chantal looked up from the flitter entrance. "That's a strange cloud formation, and the wind is picking up," she said. "I think we had better load up quickly and leave before the storm hits us."

"You're right. It doesn't look very good." This caused Mark's heart to ache to see how his ship had been ruined as they loaded up the flitter. Mark and Chantal had just finished loading the supplies when they saw a huge flash and felt thunder hit the cargo door. The shock wave caused Mark's body to be hurled back inside the flitter.

"Mark, are you alright?" Chantal cried.

Mark shook but was okay and said, "We can't leave. Try to make for the ship and wait out the storm," Once inside the ship's hull, another bolt of lightning struck the ship with a loud sound.

"Are there any earplugs in the sick bay?" Mark asked.

"Yes, I will get a couple of pairs for us." Chantal retrieved the earplugs and noticed something moving under the desk. "Misty, what are you doing here?" Misty shivered in fear. Chantal held Misty to her chest as the lightning struck the ship's hull. Chantal placed a tension bandage over Misty's head and ears as she met Mark outside the sick bay. "I found Misty, and she is frightened."

"I can't hear you. The thunder is too loud! What are your quarters like?" Mark asked.

"It is in shambles," Chantal shouted over the thunder.

"My room is not too bad, come on," said Mark as he led Chantal to his quarters. "We can spend the night here. I guess this is why no trees or animals exist on this plateau. The storms have destroyed all the plants and trees."

Chantal felt the ship shudder, and the interior lights flickered as the lightning struck.

There was no place to sit except for the bed. Chantal sat beside Mark on the bed with their backs against the ship's wall.

Chantal raised her voice to be heard above the thunder, "Why did the Thracian create so much destruction inside this ship?"

"Anger and shame," Mark replied. "The Thracians are a warrior society. To return home with one heavily damaged ship and leave another ship behind is to bring shame to their family clans. All they can do is salvage what they can; the rest is destroyed. I would like to know what these two scouts were doing here, ten light-years beyond the Buffer Zone. From what I have heard from Wayside Station, all the reported Thracian sightings are never far beyond the Buffer Zone."

As the storm continued throughout the night, Mark felt the weight of Chantal's head resting against his shoulder. Mark gently looked aside to see both Chantal and Misty sleeping. *It had been a long and tiring day,* Mark thought. Too tired to open his eyes, he closed them and gradually fell asleep.

***

It was a quiet mid-morning when Mark woke to the smell of coffee. "Good morning, Mark. You are just in time for a cup of hot coffee."

"Thanks," Mark said as he stretched his arm and took the cup from Chantal. "How did you sleep last night?" Chantal asked.

Mark replied, "I slept alright; why do you ask?"

"You were talking in your sleep," Chantal smiled.

"Oh, what did I say?" Mark asked.

Chantal replied, "Most of it was garbled. Talking in your sleep is usually stress-related."

"Is this the doctor talking?" Mark grinned.

"Perhaps." Chantal sat down beside him with a cup in her hand.

"So what does the doctor suggest about my stress?" Mark asked.

Chantal suddenly felt awkward and felt her face flush. "Excuse me," Chantal said as she stood up and exited the room. Chantal walked to the cargo door and stepped outside to breathe in the cool, fresh air.

"Melissa, what have you done to me?" Chantal whispered.

Feeling back to normal, Chantal decided the sick bay would be the next room to search. Chantal was dismayed to discover the destruction of the lab and both analyzing beds but was relieved to find the portable medical scanner from the locker intact. Chantal picked up all the bandages and antibiotic gel kits scattered on the floor. *This will have to do,* she thought, as she carried them all to the flitter. After Chantal salvaged what she could from her quarters, she entered the bridge and saw a large gray steel box on the pilot chair. The bridge was largely destroyed, with much of the room blackened around the central console and the floor uplifted by a few feet.

The gray box was attached to various cables and had a cam on top. The cam rotated toward Chantal and said in a familiar voice, "Hello, Chan. It is good to see you again!"

"Sara, you are back!" Chantal cried.

"Yes, thanks to Mark. Although I feel restricted, it's good to be back!"

"Where is Mark?" Chantal asked.

Sara sighed, "Mark went for a walk, and I don't think he will be back for a little while. He needed to cool off after I told him what had happened."

Chantal asked, fearing the worst, "What did you tell him?"

"The whole scenario here was a set up by the Commonwealth Navy authorities."

"I don't understand?" said Chantal.

Sara continued, "An hour after you and Mark first flew north for the Cambridge mine, we came across another Thracian Scout ship that was more severely damaged than the first ship we had encountered. The ship was erratic in its flight toward us and crash-landed. Fearing an explosion, we lifted the *Marvel* above the scout ship when their ship exploded beneath us. The *Marvel's* lift drive was disabled, and we crashed to the ground, killing Bill and Mason. The *Marvel* was beyond repair, but the power control was still operational."

"Then you were still functional," said Chantal.

"Yes, I could still see and hear, but with the com-link external transmitter disabled, I could hear Mark but couldn't reply."

"Several hours later, the first scout ship arrived on site. One of the two crew members checked their ship and searched *Marvel*. The Thracians were very powerful and could tear open or burn down the *Marvel's* interior doors with their laser pistol. The Thracians discovered the bodies on the bridge and connected them back to their ship. I hacked the communication device back to their ship's computer and discovered the truth. The Thracian took both Bill's and Mason's bodies and dumped them outside. The other Thracian joined them and dismantled the drive pile for their own. When the power plant went offline, I became unconscious. Mark told me the rest of what took place afterward."

"What does all this have to do with the Commonwealth?" Chantal asked.

"It's is my assumption that Rear Admiral Melnyk was distraught when he failed to haul in Mark and the crew without drawing the attention of the local authorities and public. We know far too much about the robotics and the new cyborg that was supposed to be obliterated 100 years ago. Mark knew this and blackmailed the rear admiral to buy the extra time needed for our freedom. We outsmarted the Commonwealth by staying one step ahead; however, the Commonwealth had Plan B as a backup. The Commonwealth regulates all traveling and trade routes, so they knew about our last stop here. When I hacked the Thracian ship's computer, I discovered that the twin scout received a message from an unknown source that a ship with a description like *Marvel* was unarmed and had black diamond cargo, otherwise known as the "Carbonado" onboard. It was a temptation they couldn't ignore! Carbonado is highly valued by Thracian culture and society. That's probably one of the reasons why they went into a rage when the black diamond wasn't found. They intended to have both ships use their traction beams to seize *Marvel*. If they were successful with a captured ship, crew, and the black diamond, the four Thracian crew members would have their status raised and become very wealthy! But they picked the wrong ship, the Cambridge Yacht. In desperation, the Cambridge Yacht captain tried to evade the two Thracian scouts and fired on them with a laser cannon. The laser hit one of the scout's drive piles at close range. Before the drive pile created an uncontrolled chain reaction, it was ejected from the ship and then exploded. The explosion damaged the Cambridge Yacht and both scout ships. Mark told me that the Cambridge flitter had left before the mother ship exploded. The scout that lost its drive pile stayed behind, and the other ship landed immediately. The scout located the Cambridge mine site and destroyed both flitters and the mine site. I assume that the two crew members were hurt quite badly, which caused them to crash their ship near the *Marvel*."

"The Cambridge flitter had a young girl with them," Chantal whispered.

"I didn't know that. Mark didn't mention the girl. Chan, what's left of my remaining sensors shows that Mark is walking back to this ship. Knowing Mark, he likely has something up his sleeve."

***

Chantal watched Mark modify one of the four egg-shaped escape pods. Each was equipped with a single seated electric drive unit designed for safe passage to the nearest planet within a solar system.

"How are you doing, Mark?" Chantal asked.

"I am almost finished with the third escape pod. Removing the seat and protective equipment allowed me to make room for another power plant from pod number four. This should double the pod's speed to half the speed of light. Sara has reprogramed the pod's computer to aim it toward the nearest major commercial solar system at 18 light-years."

"That's still 36 years away," said Chantal. "Are you sure this is the right thing to do?" Chantal asked.

Mark replied, "Chan, the Commonwealth used us, directly and indirectly, and have committed murder. When the pod enters the solar system, our contract, video, and story will be released on the public communication network. Perhaps those who are responsible will be held accountable."

"Or perhaps not!" said Chantal.

"Or perhaps not," Mark said dryly as he nodded. Alright, Sara, you can take over now." Mark stepped back and watched the pod's top-half glass canopy slide forward, lower, and lock to the bottom half. The pod rose noiselessly and vanished into the atmosphere.

"Thirty-six years is a long time," said Chantal.

Mark put his arm around her shoulder. "At least we have each other for company in the meantime. Chantal felt her heart beating and, for the first time, didn't mind.

***

After installing the fifth cam outside the homestead, Mark asked Sara, "Can you see the Vica village and the valley?"

"Yes, thank you," Sara replied.

"Good, I am hungry. Have you made lunch yet?"

"Bad joke, Mark!"

"Not having Bill around has made a huge adjustment in my cooking habits," Mark sighed.

"You and Bill were good companions for over 50 years," said Sara.

"Which makes it so hard to see Bill gone," said Mark. "All the more reason I am glad to have Chan for company now. I have taken your advice to let Chan make the first move to initiate a relationship, but it's been three and a half months since the accident."

"Her body is still changing, which is giving her mixed feelings about how to sort out her emotions. Right now, Chan's worst fear is that you may reject her. Chan still believes she is a freak."

"So, how do I get around that without telling Chan the truth about her?" Mark asked.

"When the right opportunity comes, you will know," said Sara. "Is there something else bothering you?" Sara asked.

"The coming Commonwealth and the Thracian war may prevent the Cambridge company from sending a rescue party for us. And there is this problem of maybe 36 years or more before we are rescued and having only eight years of food to sustain us. The Cambridge Brothers told me that they hadn't found any local food that isn't harmful for human consumption."

"We have one major advantage over the Cambridge Brothers," said Sara.

"What is that?" Mark said, looking surprised.

"Chan is a qualified medical doctor and was originally trained as a bio-researcher. We have eight years to find the answer to your predicament. And she found an undamaged medical lab scanner in the sick bay locker.”

“I hope you’re right, Sara.”

***

Tired from his daily constructions on the cobblestone road, Mark met Chantal at the healer’s house.

“How is Doctor Chan coming along with Marica, the healer?” Mark asked.

“We shared a lot about anatomy today. They share a lot of similarities with Homo sapiens. They have a superior digestion system, which may explain why they can consume crops that would harm us.”

“Wait a moment, Chan; there is Asoma, and I need to talk with him.” When Mark was a few feet away from Asoma, he heard a loud, piercing cry from a Zaroma woman standing outside her doorway several houses up the street.

Asoma and the other villagers bowed their heads silently.

Chantal whispered in Mark’s ear, “Maybe we should do the same.”

Mark nodded his head in agreement. One by one the local villagers approached the silent woman bowed their heads and walked away. Mark asked Asoma, “What just happened?”

Asoma replied, “There has been a death, a mate of hers. We bow our heads to show respect.”

"Thanks, Asoma. It is alright for us to show our respect?" Asoma, surprised, looked at Mark and nodded his head.

Mark and Chantal walked up to the widow, bowed their heads and continued back to their homestead.

***

"Well, that was a good meal. How would you like to join me outside on the lookout bench before it gets too dark," Mark asked.

"Give me a few minutes, and I will bring out the coffee."

Mark walked a short distance to the hill's edge and sat on a twig bench.

For many shipwrecks, we are not doing too bad, Mark thought.

"Here is your coffee." Chantal passed the cup to Mark and sat beside him.

Mark enjoyed the warmth of Chantal next to him and felt sure Chantal's feelings were mutual.

"I have been thinking about that older woman's mate who died today. Do you think the Zaromas have a soul?" Chantal inquired.

"Yes, I believe the Zaromas do have a soul. Asoma and I discussed the Creator and afterlife, so there is some common ground here. I asked Asoma where his people come from, and he said he doesn't know much about his ancestry. What has puzzled me is that there is an old myth about the old world. This valley and another valley around it were empty of Zaromas for many long-ago generations. To live a peaceful communal lifestyle, they arrived here to escape the war and destruction in the north before the ice age. As far as I understand, they have achieved this. All the nearby villages in other valleys kept their population under 500. They are advanced for their stone age; they showed me how to construct short hollow blocks out of the stone and align them together to make a water pipe. Much like we did during Roman times.”

Chantal struggled to force herself to ask Mark, “May I ask a personal question about you and Katherine?”

Mark turned sideways toward Chantal, “Sure, what is it?”

“When you first fell in love with Katherine, did that love make sense to you both?” Chantal asked.

Mark replied, “I asked a similar question to my Nana when I was 15 years old about a girl with whom I had fallen in love.” She said, “Love is an absence of judgment.”

“That was your Nana’s answer?” Chantal cried.

Mark laughed, “Yes, but she added more, though.” Mark reached over and put his arm around Chantal’s shoulder.

“I was only 15 years old, too young for a serious relationship. Nana also said that when I am older and meet a girl who wants to share my dreams and vice versa, and there is love between us, then love will begin to make sense.”

"That's one smart Android," Chantal whispered.

"Yes, she was," Mark whispered. Mark kissed Chantal's lips and, seeing no adverse reaction, kissed her again. Chantal moved closer to Mark and passionately kissed him.

In an uncontrolled panic, Chantal stood up and returned to the house. When Chantal closed the door, she walked several circles inside, banging her fist on her upper thigh.

"Sara, this isn't working for Mark and me!" Chantal said, frustrated by her emotions.

"Aren't you in love with Mark?" Sara asked.

"Yes, but I am just a freak, and he still thinks I am an android! How can anyone be in love with a woman like me?"

"Do you know what you just said?" Sara asked.

"What do you mean?" Chantal replied.

"You just called yourself a woman. When was the last time you called yourself a woman?" Sara asked.

"I don't know... I am not sure. I am going to bed. Goodnight!" Chantal, feeling confused, walked to her bedroom and slid the curtain closed. Chantal sat on her bed with her knees tight against her chest and silently cried.

***

All the village dwellers sat on the courtyard's cobblestone, waiting for the wedding ceremony to start. The courtyard

267

was lined with benches outside, where Mark and Chantal sat together, holding hands.

Misty mingled among the children whose parents had now accepted her presence.

"Here comes the bride and groom." Chantal whispered in Mark's ear, "The bride and groom are dressed in purple!"

The bride and groom walked to the central part of the court, where the five elders awaited them. Each of the elders took turns speaking to the engaged couple.

"What are they saying?" Chantal asked.

"I can hear only some of the words. Something about the wisdom of life and marriage," Mark replied.

When the five elders finished, the couple exchanged arm bracelets. The married couple turned around to exit the court.

"That was nice. Simple and no fanfare... Mark?"

"I am right behind you." Chantal stood up and turned around to see Mark on one knee.

"Mark, what are you doing?"

"It has been almost five months since we were shipwrecked. I have learned to love you with my entire heart and mind. I want to share my life with you, and I want to be part of your life. Chan, will you marry me?"

"Mark!" Chantal shouted. "You can't marry me! You don't know who I am. I am just a freak! One huge bloody freak!"

Mark stood up and stepped over the bench to face Chantal. Mark spoke softly, "Beta 6, otherwise known as Astern. I like Chantal better," Mark smiled.

Chantal replied in the same soft voice, "Has Sara told you about me?"

"No, I was suspicious when Lee the android scanned your right neck instead of the left side. I felt there was a serious security breach. At the ship, I overrode Sara and looked up what she had on your profile. That is when I became aware of your true identity. Sara and I thought it was best to wait until you trusted me enough to reveal yourself. Melissa also told me what she had prepared for you back at her ship. Even if Melissa hadn't, it wouldn't change anything; I still want you to be my wife. What the Kazakhstani genetic researchers did to your embryonic development was a terrible thing, but you are still no less a human being than I am." Mark added, "Will you marry me?"

Chantal's tears slowly ran down her cheeks. She nodded her head and replied, "Yes." She embraced Mark and rested her head on Mark's chest. Moments later, Chantal asked in a gentle voice, "Mark, is everyone watching us?"

"All 350 of them, with Asoma just behind you." Mark smiled.

Chantal took a deep breath and turned around to face Asoma.

"Is your ritual finished?" Asoma asked. Chantal nearly laughed in embarrassment but caught herself just in time.

Mark intervened, "Chan agreed to be my mate. We would be honored if you and the other elders would perform our communion."

"Come with me," Asoma said as he led them back to the middle of the court.

"Mark, what about a ring?"

"I have my mother's gold ring with me."

"But Mark, I don't have a ring for you!" Chantal said with a hurt look.

"Asoma has Melissa's ring. At the right time, he will pass the ring to you."

Chantal couldn't believe what was happening. For the first time in Chantal's life, at 84 years of age, she felt like a true woman and was about to marry a man she dearly loved.

***

In the bedroom's darkness, Chantal's voice called out, "Mark, did Katherine have your mother's ring, too?"

"No," Mark replied, "The initials didn't match her name."

Chantal turned on the light and looked at the gold ring. "It says C.A.S. What was your mother's name?"

"Corrie Aileen Styler. Now the ring says Chantal Astern Styler. So you see, Chan, God's unseen hand still works for us."

Chantal smiled, placed the gold ring back on her finger and turned off the light.

Chapter 14 - Year One – Honeymoon

Twenty feet above the Vica river, two bold red labels that read Escape Pod Number 1 and 2 raced each other as they followed the river south downstream. Mark, in Pod Number 1, quickly joined Chantal's pod as they flew along the narrow River Canyon. A 1000-foot high green jade cliff came into view as the watercourse made a sharp circle at the base. Both pods stopped to appreciate the grand view of the humongous cliff of jade.

"Mark, that mountain is just gorgeous!" Chantal cried through the communications channel.

"We can go over or around it," said Mark.

"Let's go around," Chantal replied as they continued along the canyon.

"Have you been here before?" Chantal asked.

"Not at this elevation, just at a 10,000-foot level," Mark replied.

The pods approached a 200-foot high waterfall gracefully angled downwards over the fall and to the river below. The semi-tropical rainforest replaced the dryer temperate forest to the north as it flowed south to the ocean.

"The climate here is getting warmer, and we are almost at our destination," said Mark. He led Chantal to the right side of the river at the base of a small twin waterfall. Mark turned his pod to the right before the fall, angled up a couple of hundred feet, and entered through a narrow opening into the mountainside. On the other side was a significant bowl-like depression with a small lake.

"It is a beautiful setting, but what is so special about it?" Chantal asked.

"Asoma said cold glacial water feeds most rivers and lakes. A hot spring feeds a small creek that runs into the shallow lake, and there are no predators, so it is safe for us to swim in. The bowl is sheltered from the major winds, so the clearing near the lake should be a good spot to land the pods."

"I love it, Mark. Let's do it!" Chantal raced ahead and landed on the level ground with Mark's pod a short distance away. The pod's glass canopy popped up and slid back. Chantal quickly climbed out and ran to the lake. She dipped her hand in the water and said, "There must be another hot spring on the bottom of this lake; the water is warm."

"Help me set up the tent on the soft ground over here, then we can go for a swim," said Mark.

"Are we alone?" Chantal asked.

"The pod's scanner doesn't register any Zaroma natives. But there is a fire pit near the hot spring," said Mark, adding, "If there is a fire pit, then there has to be a trail nearby, too. It might be a good way to go exploring. Maybe we will find something interesting."

"I can't wait anymore. The last one in the water gets to make supper." Chantal quickly stripped and jumped in the lake. Mark grinned and heard Chantal shout, "The water is fantastic. Hurry up!"

Mark stripped down and dove into the lake. Chantal stood on the lake bottom, the water up to her neck, waiting for Mark to break the surface. Chantal screamed as she felt something touch her belly button and pushed away.

Mark's head rose through the water, grinning. "Mark! You are disgusting!" Then Chantal laughed, put her arms around Mark's neck and said, "This is going to be a perfect honeymoon. I love you!" Chantal pulled Mark in with her arms and passionately kissed his lips.

***

"Are you ready to start hiking again?" Mark asked.

"Yes," Chantal replied, "Let's go west toward the Twin Falls. I want to find out where that trail will lead.

Chantal noticed that the entire area seemed to be the remains of an old volcano. The balsamic rock indicated a volcano, a geological rarity among the serpentine and jade environment. *That narrow split when we came in on the west side of this crater*, Chantal thought, *drained the lake overflow*. Mark and Chantal hiked along the trail through the split mountain, which eventually turned south and stopped at a high cliff.

"Why did the trail stop here?" Mark wondered.

"Look up there, about 20 feet up," Chantal pointed.

Mark stepped back and looked at where Chantal was pointing. It was a tunnel. "This is getting interesting; why is there a hole in the rock wall?"

"If we follow the ledge on the cliff face, it seems to work its way up to the tunnel. Wait here, and I will check it out," said Mark.

"Not very likely, where you go, I will go too," said Chantal.

Mark thought it was better not to argue and took hold of Chantal's hand as they walked up the narrow ledge, which angled up to the right and then to the left. When they arrived at the tunnel, Mark pulled out his light torch and set it on a high beam. The tunnel was a lava tube. Mark scratched the inside wall surface with his fingernails. It looked like soft white gypsum, but no gypsum could be seen outside. The rain had likely washed it away, probably thousands of years ago. The lava tube was small, just high enough to stand up.

"The ceiling is black with soot. The Zaromas must have used a burning torch to see their way in this tunnel. Shall we see where this tunnel will lead to?" Mark asked.

"Just don't let go of my hand. Otherwise, you will have a burst ear drum!" Chantal warned.

Chantal had mixed feelings of excitement and eeriness while walking through the long, dark, winding tunnel and kept a tight grip on Mark's hand. The tunnel had several main branches, but it was easy to stay on course because of the black soot covering the ceiling. Eventually, she could see the light ahead at the end of the tunnel. Chantal emerged from the tunnel several hundred feet from the valley floor. An open coastal valley, River Delta, and ocean could be seen from several miles away, but the thick forest obstructed much of the scenery.

"The trail seems to lead near that cliff; we should get a better view without the trees blocking our view," Mark said.

The narrow trail cut through the thick, semi-tropical, fan-shaped plants and trees as Mark and Chantal followed the path down the cliff. A loud crack was heard overhead, and something alive fell on top of Mark, causing him to hit the ground. The creature got up on its feet and shuffled down the trail.

"Are you alright?" Chantal asked.

"I think so. What was it?" Mark said as he rubbed the back of his neck.

Chantal replied, "It was a young Zaroma child. I think the poor thing is hurt. There are spots of blood on the ground."

Mark and Chantal continued along the beaten path until they came to the opening cliff. A male child was standing near the cliff's edge, holding his right arm and trembling with fear.

"It is a boy, and his arm is cut!" said Chantal. "What is he doing here all alone?" Chantal asked.

"Maybe he is lost. He is probably hungry," Mark replied.

Chantal removed a protein bar from her shirt pocket and unwrapped it.

"Won't that bar harm the child?" Mark asked.

"The Zaromas can eat many of our foods, but we can't eat theirs." Chantal spoke with the child in Zaroma, "We won't harm you. Are you hungry? You can eat this if you like."

The child, frightened by the two strange creatures he had never seen before, moved closer to the cliff.

"Look!" Chantal said as she sat down near the child, took a piece of the bar and stretched her arm back to the child again. The hunger pain was too much for the child to ignore, he reached out, took the bar and ate it in one mouthful.

"Where are you from?" Chantal asked. The child pointed toward the village by the ocean shore.

Mark noticed a thin line of smoke by the ocean shoreline that he thought must be the village of Mica. "Is that village called Mica?" The child didn't answer, as he was unsure if he could trust them.

"We are from the village of Vica," said Chantal.

"Are you the Star People?" the child asked.

"Yes, but Vica is our home now. Would you like another one of the bars I just gave you?"

Mark removed the wrapping from the protein bar and handed it to the child.

"My name is Chan, and my mate is Mark. What is your name?"

"Kogi," the child answered.

"What are you doing here all alone," Chantal asked.

Kogi replied, "It was near dark, and a barbed tongue chased me. I got lost in the darkness and couldn't find my way back. I climbed the tree, then the branch broke, and I fell on Mark."

"Kogi, your arm is cut. I am a healer, but I can't fix it here. My herbs are back at our camp. Mark, can you help me?"

Mark sat beside Chantal, "I think your people call it the Place of Tranquility."

"I know it," said Kogi.

"There is a tunnel nearby. Will you hold my hand? We will take you to our camp so that I can heal your arm. Then we can take you back to your home. Would you come with us?"

Kogi nodded his head. "Mark, will you please lead the way," Chantal said as she held Kogi's hand.

***

"How is your arm feeling now?" Chantal asked.

Kogi looked at the bandaged arm and said, "You are a good healer."

"We are all packed and ready to go," said Mark.

"Do you understand how we are taking you home, Kogi?" Chantal asked.

"We are going to fly like a bird to my village," said Kogi.

"Chan, it will be a tight fit with Kogi on your lap. Our pod's computers are systematically connected so that we will fly together as one unit," said Mark.

Chantal boarded Pod One, and Mark lowered Kogi onto Chantal's lap. The glass canopy slid forward and lowered itself down over their heads.

"I am going to put my arms around you to protect you," said Chantal.

Mark sat in his pod and asked Chantal over the com-link, "How is Kogi doing?"

"So far, he is doing fine," Chantal replied.

Mark operated his pod as they rose together into the atmosphere. The pods flew over the old volcano toward Mica.

The pods circled the village in a few short minutes and landed near the fishing pier. Mark stepped out of his pod to help lift Kogi and Chantal out. Kogi walked to the elders near the crowd and talked with them for several minutes before they all returned to their homes.

Mark grinned and said, "Well now, I guess this is our thank you for bringing Kogi home."

"Before we leave, Mark, I would like to take some tissue samples from the fish the Mica fishermen caught." Chantal grabbed her kit from the pod and walked to the cannery by the pier. Chantal was amazed by the variety of fish hung up to dry or laid in salt containers. Chantal began taking samples from many small and sizeable strange-looking fish.

Mark stood outside the cannery, observing the area, and saw a small group approaching. He said, "We have company."

Kogi and two adults separated themselves from the elders. Kogi held an object in his hands as he bowed. He said, "This is my mother and father," who bowed.

Chantal and Mark returned their respective gesture.

Kogi's father spoke, "Our elders approved our token for bringing our son home."

Kogi gave Mark a pure white jade bowl.

"We thank you for the gift. It is a great honor," said Mark.

Chantal turned toward Kogi's mother, "May I hug your son for the last time?" Kogi's mother looked at her mate for understanding, and he nodded.

Chantal fell to her knees to hug Kogi and immediately returned to her pod.

"Did we offend your mate?" Kogi's father asked.

Mark sighed and said, "No. We don't know if we will have a child of our own. Thank you again." Mark turned around and walked to his pod.

Both pods flew above the mountaintop back to Vica's homestead as neither spoke on their way home.

It was late at night, as Mark and Chantal couldn't sleep when Mark felt Chantal's tears on his shoulder.

"Chan, I believed Melissa when she said you would have all of a woman's reproductive organs."

"And when will that be? I am 84, and you are 81. We are not getting any younger," said Chantal.

"If it's any consolation, my father was 90 when I was born. And our bodies can keep their youth until our threshold at 120! So we still have a lot of time. Give Melissa a little more time to prove she was right," Mark suggested.

Chantal paused for a few moments to kiss Mark and tightly hugged him.

Chapter 15 - Year One to Two - Cheri

Taking a break from working on his community garden, Mark leaned on the hoe to rest his back as Asoma walked toward him.

"Good morning, Asoma," said Mark.

"Good day to you. We are having an interesting event today."

Whenever Asoma says, 'Interesting,' it is something important! Mark thought.

"What is it?" Mark asked.

"Come and see," Asoma replied.

Mark dropped his hoe and walked back to the village with Asoma. When they arrived at the village court, Mark was stunned to see Chantal with a young human girl dressed in local gray clothes several sizes too large sitting on her lap.

"Where did she come from?" Mark asked.

Chantal, astonished, said, "I don't know yet. She is in shock! She hasn't said anything or responded to any verbal commands."

Mark asked Asoma, "What do you know about this child?"

Asoma replied, "From the village, Paca, across the Nica river, west of the Cambridge mine." Asoma continued, "The child was found wandering nearby six months ago. We believe the child has a sickness in her mind. When the Paca heard about the Star People living among us, they brought her here. The child would not eat, and they had to force her to eat. But much of our food doesn't stay down."

"She is all skin and bone. We have to get her home and feed her," said Chantal as she lifted her up in her arms and walked back to their homestead.

"Thank you, Asoma. We will take her home now," said Mark.

***

Chantal sat on the floor with the child for an hour. After growing tired and frustrated, she stood up and walked to the table. "I can't get her to eat anything. If I force her to eat, she either spits it out or chokes on it."

Mark pondered, then asked, "Any suggestions, Sara?"

"Every child likes something sweet. Moistening the dry apple sauce with warm water might work," Sara replied.

Mark took the freeze-dried apple sauce from the storage room and opened the package. Then he mixed some apple sauce with warm water with the spoon but still could not get the child to eat. Mark left the bowl on the floor to sit at the table.

"I think the poor child must have witnessed both her mother and father's horrible death. It's bad enough for an adult to witness the 'G-Gun' destruction, but for a child?" said Mark.

Chantal said in a surprised whisper, "Mark, look!"

Mark watched Misty with the doll in her mouth, gently rubbing it against the child's face.

"I forgot all about the doll we found near the mine site; it must have belonged to her," Chantal whispered.

After a few minutes of Misty trying to get the child's attention, her eyes began to focus on the doll. The child slowly reached for the doll and studied it in recognition. Then she held the doll to her chest.

Misty grabbed the end of the spoon in her mouth, took a scoop of applesauce and held it in front of the child's mouth. The child stared at Misty, then at the spoon, opened her mouth and swallowed the applesauce.

Chantal, with one hand to her mouth and the other on Mark's hand, watched the miracle take place. Misty continued to refill the spoon and feed the child.

"I will try to mix some bananas with applesauce next," Chantal said as she walked into the storage room. Mark watched from his chair with Misty on his lap as she took over feeding the child.

***

Unable to sleep, Chantal stared at the reflection overhead of the ship clock that Mark had salvaged.

"Mark, are you awake?"

"I was," Mark said, half asleep.

"It has been several weeks now, and she hasn't said a single word yet."

"She who?" Mark asked.

Chantal slapped Mark's chest with her open fist. "Are you awake now?"

"Yep!" Mark said, gasping.

"Good, I am talking about that child in the other room! I don't even know her name."

"There is no record back at the flitter of the girl's identity," said Mark.

"Sara mentioned that when the child snaps out of shock, we will need to be there for her to help her recover mentally," said Chantal.

"How are we going to know when this happens?" Mark asked.

A high-pitched scream filled the house. "I think this may be it!" Chantal rushed out of the bedroom. The girl was sitting upright in her bed, screaming hysterically. Chantal picked up the child and held her tightly. The child painfully vomited and intermittently coughed before she quieted down into a soft cry.

"Mark, get a towel, please." Chantal's maternal instincts overtook, "It's okay, Mommy is here!"

Mark came back with the towels and wiped up the child's vomit. He sighed and then asked, "Do you know any children's songs?"

"It's been a long time, but I should be able to sing a few. Please get rid of this towel and get another fresh towel with some warm water to clean her up," said Chantal. When the child finally settled down, Chantal sang several nursery rhymes. After Chantal had finished the nursery rhymes, the child pushed away from Chantal and looked up at her face.

"You are not my mother! You are an android! Where are my mother and father?" the child demanded.

Not sure how to explain to a child as young as this one, Chantal decided to tell the truth.

"I think you saw what happened to your mother and father?" The girl silently looked down.

Chantal tried a different approach, "I am exhausted, so if you like, we can sleep together for the night."

"Okay," the girl replied reluctantly.

"What is your name?" Chantal asked, but she didn't answer the question. Chantal lifted the blankets over the both of them. Chantal turned the lights out when the child said,

"Cheri."

"What did you say?" Chantal asked.

"My name is Cheri."

"Cheri, that is a pretty name. Good night, Cheri."

Mark stepped into the bedroom and saw them both sleeping, smiled and quietly left.

***

The sunlight beamed through the bedroom entrance curtain, and Mark awoke when he heard a voice, "Move over."

Chantal snuggled up to Mark, and he asked, "How is she?"

Chantal replied, "Her name is Cheri, and she is sleeping like a baby for the first time. I would still like to know how Cheri walked 20 miles across the Water Stone and Nica River in the right direction to the nearest village?"

"Maybe God had a hand directing Cheri's walking," said Mark.

"I am not sure if there is a God yet. I am going to make some porridge for Cheri. Do you want some?"

"Sure!" Mark replied.

Chantal got up from the bed into a change of clothes and walked into the kitchen.

Cheri, Mark thought, *how did you manage to survive to this day? I probably will never know. She looks about three or four years old, a year too early to have an implant. Otherwise, I could have uploaded her data and learned about her journey. Oh, well, it's time to get up.*

Mark rose from his bed, dressed, and stepped outside to the outhouse. Mark saw Cheri alone on the deck rocking chair when he returned to the house.

"Mind if I join you?" Mark said as he sat in the other rocking chair.

Finally, Cheri spoke, "Why did they kill my mommy and daddy?"

Mark sighed, *Darn it! I knew she would ask that question!* Then, a thought came to Mark, and he replied, "Sometimes we don't know why some people get to live, and some have to die. My mother died when I was about your age. All I had left was my father, and I didn't see him much. But I had a devoted Nanny."

"Was your Nanny an android like Chantal?" Cheri asked.

Mark replied, "My Nanny was an android, but Chan is human like you and me."

"How do you know Chan is human? She has gray skin like an android," Cheri asked.

"People come in different colors: white, brown, black, and sometimes gray. Tell me, Cheri, what do you know about androids?"

Cheri considered the question and then said, "They are not real people. They can't have babies, they are not born like babies, and they don't have red blood."

"Very good, Cheri, you are a very smart girl," Mark smiled.

Chantal popped her head through the door and said, "Breakfast is ready."

"Just a moment, Chan," Mark said as he entered the storage room. Chantal and Cheri sat around the kitchen table, waiting for Mark. Moments later, Mark sat across from Chantal and removed a needle from the medical kit.

"Chan, please do a demonstration with this needle for Cheri to see."

Chantal looked at Mark for a few seconds, then took the needle and poked the back of her hand.

Cheri was able to see red blood seeping out of her hand. Mark handed Chantal a small cloth to wipe the blood away.

"Now, may we say our grace?" Mark asked.

***

Chantal waited outside the nursery day care for Cheri to finish her day. Misty heard Cheri's footsteps, and ran to meet her at the door. The door opened to allow Cheri and all the youngsters outside.

"Hi, Misty," Cheri patted her head. Cheri reached for Chantal's hand.

"Where are we going today?" Cheri asked.

"We are going to meet up with Mark down at the riverside for a picnic supper," Chantal replied.

"How come we don't eat with the others?" Cheri asked.

"Something in their food will slowly harm us," said Chantal.

"Like when I was sick all the time?" Cheri asked.

"Yes, but the poison is gone from your body now. That is why you are feeling better now."

Chantal and Cheri crossed the footbridge to the other side of the river and walked down 50 feet before it leveled out to the multi-colored pebble riverbed. At the high end of the riverbed were several green stone benches.

"There is Mark!" Cheri cried as she rushed ahead of Chantal.

Mark bent down to pick up Cheri and hugged her.

"Can we go swimming?" Cheri asked.

"Sorry, Cheri. It is too dangerous to swim in the river. The river is too deep, and the fish in the water will eat us. Watch this," Mark said as he picked up a fungus and threw it in the middle.

A large silver-lined green fish larger than a man broke the surface, swallowed the fungus in one bite and dove back into the river.

"Are you guys going to eat or not?" Chantal shouted.

"Let's race back to the benches," said Cheri as she and Misty ran ahead of Mark while he grinned.

After eating, Mark and Chantal sat alone while Cheri looked for a shiny pebble to take home.

"Cheri seems well-adjusted to her new surroundings lately," Mark said quietly.

"It has been three months here for Cheri, and her nightmares are very few now," said Chantal.

Mark nodded in agreement, "I am very pleased that she got over it. My nightmares over Kate's death were bad enough for me."

Chantal pressed her lips, then whispered in Mark's ear, "I have a surprise for you."

"What kind of surprise?" Mark asked.

With a tight smile, Chantal took Mark's hand and placed it on her tummy. Mark initially was puzzled by Chantal's gesture and then looked at her face, "You're kidding me!"

"I kid you not!" Chantal said with a wide grin. "I don't understand how it is possible, but Melissa also gave me a new reproductive system. And you are going to be a father."

Mark was speechless.

"The scanner registered a boy," said Chantal, and added, "Is it alright if we name him Jonathon?"

Mark grinned, then replied, "Jonathon is what I would like to name him, too."

"Hey, look!" Cheri pointed at the sky.

Chantal looked at the river downstream; a huge waving dotted cloud was moving upstream. It soon became apparent that thousands of small yellow and blue birds with short wings and beaks were flying low above the river. The line of birds seemed endless as they flew low over their heads as they continued upstream.

"Do you think this is an omen, Mark?" Chantal asked.

"Omen or not, it's time to leave quickly. The clouds are almost black; the storm is coming." Mark shouted, "Everybody, let's head home now."

The neew Styler family all laughed with excitement as they hurried home.

Chapter 16 - Year Eight - Famine

Jonathon pointed to a round purple object covered with short, sharp prickles. "What is that?" Jonathon asked.

"Unknown rotten fruit," Chantal replied.

"Maybe that one will save us," Jonathon smiled.

Chantal said, annoyed, "I thought you said you were tired?"

"I am!" Jonathon said as he rushed to bed.

Chantal stared at the partially rotten fruit on the table. She sighed and carefully picked it up because of the spiked skin. The scanner registered the alkaloid on the fruit skin. She cut the fruit in half and scooped out the soft interior paste. She analyzed the fruit with the scanner, and her mouth fell wide open.

Chantal banged on the door late at night, "Who is it?" Marica shouted from inside.

"It is me, Chantal. Please open the door!"

The odors from the healer's herbs nearly overwhelmed Chantal when Marica opened the door.

"Marica, what is this fruit?"

Marica picked up the remains of the cut-up fruit from Chantal's hand. As her eyes focused on the fruit, she recognized it. "It is a muster fruit from a desert plant east of here. There are many fruits similar to those of other muster plants. We call the desert plant muster because we use it to prepare for battle against the barbed tongue. We peel the skin with the spikes and wrap it around a bola rock to fight against a pack of barbed tongues. The spikes severely agitate their bodies and can cause panic among them and make them disperse back into the forest. Is this fruit safe for your family?" Marica asked.

"Yes, Marica! This fruit has most of the nutrition we need," Chantal replied. "Could Kon take us to where the muster fruit grows?"

"Kon should be able to help you. Ask Asoma first for his approval. Sit in the morning elder meeting and wait for your turn to speak. I am sure Asoma will speak about your request."

"Thank you, Marica. I will be there," said Chantal, excited with her discovery.

***

"Kon, are you sure you can direct the Styler's flitter to the desert canyon? You are only an intern scout," Asoma asked.

"My master scout has confidence. I can guide the Stylers safely back to Vica," Kon replied.

"I have told our friends about the dangers of their journey," said Asoma. "I fear the Stylers are overconfident in facing the danger. Be aware of the threats that might endanger them," Asoma warned.

"I will, and don't worry about them; they are my friends, too," said Kon.

Asoma and Kon turned their heads when they heard the flitter from the Stylers' homestead. The craft lifted up in the sky and moments later landed in the open field nearby. Kon entered through the open side door of the cargo haul and the rear door to the cockpit of the flitter.

"Welcome aboard, Kon," Mark announced. You can sit in the third seat behind us. We will be taking our flight slowly to conserve fuel, so it will take about one hour to travel."

"One hour, not a ten-day journey?" Kon said, astonished.

"Yes, but our days in this flitter will someday be over when we are out of fuel," said Mark.

Kon sat in the chair and stared out the window that had been missing. "We can see out, but not in from outside this craft?"

"I know this is hard for you to understand, but this is a one-way steel glass window. The regular glass windows on some of the Vica buildings are too fragile for this flitter," said Chantal. "We are going to follow the road east to Nica. From there, we need you to give us directions and avoid any possible dangers."

"Are you ready?" Mark asked.

Kon, more excited than afraid, nodded his head. He took a deep breath as the flitter slowly took to the sky. Kon was amazed by the sights below and wondered if things would ever be the same again. The flitter followed the road to the village of Nica, a two-day walk, in only a 20 minutes via flight. Nica was located just beyond the eastern foothill of the Nica Valley.

"Where do we go from here?" Mark asked Kon.

"There are four trails east of Nica across the Nica River. One of the trails travels south and then goes up along the foothill of the Great Divide and over the mountain pass. On the other side of the mountain trench is another village, 'Whiz'. The people there are very unfriendly."

"Why are they unfriendly?" Chantal asked.

Kon replied, "The Whiz worship an unknown deity. They trust no one outside their village. All trades are negotiated at the summit, like your muster fruit."

"Is this where we can find the muster fruit?" Chantal asked.

"No, it is further on. Before the trail arrives at Whiz, the path splits into two. The trail on the right continues to pass Whiz and over the next valley, then turns south. The trail leads to the west end of the desert. Eventually, the trail leads to the Great Depression, where we will find your fruit."

"Alight, I am taking the flitter up along the mountain trail," said Mark.

Mark maneuvered the flitter along the trail halfway up the high mountain before the path curved around the side. The trail continued east between the two peaks before angling down the hill to the next valley.

"It is much dryer here," said Mark. "That must be Whiz down there."

Chantal asked Mark, "The scanner registered a large structure further to the northeast behind Whiz. Do you want to check it out?"

Mark replied, "No, the muster fruit is our priority. We can check it out later, on the way back home."

"Do you know what this structure is, Kon?" Chantal asked.

"It is mystifying news to me. None of us are allowed near Whiz."

"That settles it; I am taking us on the other trail," said Mark. The flitter followed the trail eastward to the next valley and almost immediately turned south before turning east again. The land formation changed to an open, flat, rocky desert.

"Nothing is growing here. The scanner shows that the geology lacks jade and serpentine," Chantal announced.

"It's hard to see the trail from up here," said Mark. "I am going to lower the flitter." Mark lowered the flitter until he could see the trail much more clearly.

"We won't last very long outside; it's 114 degrees Fahrenheit outside," said Chantal.

"How much further do we go, Kon?" Mark asked.

"It's a half-day walk, so we should be only a few more minutes, and we will arrive at the depression."

"I think Kon is correct; the scanner shows humidity is increasing," said Chantal.

"Yes, there!" Kon pointed. The desert floor opened into a vast circular depression a 1000 feet deep and 10 miles wide before narrowing down south to a canyon.

On the northeast of the depression, another shadow canyon, about 200 feet from the surface came into sight. A waterfall from the canyon flowed into the depression and plunged 800 feet into the lake. The water from the fall created a mist and microclimate humidity for the valley. The lake water emptied into the river and exited at the southern canyon. Many plants and wildlife were present near the lake but became sparser further away.

"Any evidence of intelligent life?" Mark asked.

Chantal replied, "No, the scanner showed nothing larger than a small animal."

"My master scout told me the barbed tongue is the only danger we need to be concerned about," said Kon as he continued. "Your muster plants are near the big river bend in the middle of the depression! The muster plants come in many shades of colors and sizes."

"That river bend does look like a good place to land." Mark lowered the craft between the flock of birds as they took off in the air and settled it gently on the hard-packed gravel near the river.

Chantal exited the flitter and walked to the muster plants. She noticed they resembled cacti from Earth, but the fruits hung from the main stem.

"It looks like Marica is right; the muster plants come in different varieties. Kon, while Mark and I analyze the different muster fruits, could you scout upstream and see if any other plants may interest us?"

"That's part of my training. I observe plants as a scout. I will be back before sunset. Could Misty come with me?" Kon asked.

"Misty is getting up in age, Kon, so please be careful with her," Chantal warned.

"I will be careful with her," said Kon.

Chantal watched Kon and Misty wander upstream, but they were soon out of sight.

When the coast was clear, Mark asked on the com-link, "Sara, are you with us?" Mark asked.

"I am," Sara replied. "It looks like you have a big job collecting and analyzing your muster fruits. Salvaging the dryer and processor from the *Marvel* might finally come to good use. But I recommend you remove no more than two-thirds of the fruits from each plant. We need to make sure there is plenty left for the local habitat. Also, we should analyze the soil here. I assume that the cyanide is absent in the soil here."

"While you are doing that, Chan, I will scout downstream," said Mark. "I won't be long, maybe a couple of hours."

Chantal, alone, stepped out of the flitter and stared at the surroundings. She thought that it was like another part of the world here. The soil was brownish, not like the various colors of serpentine and jade found elsewhere. Chantal walked toward the muster plants and saw the short, thick needles on the main stem. Some of the muster plants were short, a couple of feet high, while others were twice her height.

"The fruit of life," Chantal heard herself speak aloud. *Why did I say that?* Chantal asked herself. Chantal's eyes settled on the nearly one-thousand-foot cliff in the distance.

There was no sign of jade or serpentine. There was just the regular brown soil and grayish gravel under the flitter. The mist from the waterfall seemed to provide a humid environment. Chantal thought, *The humidity might make it quite damp at night.* The different sizes and types of birds were found nowhere else. Some were larger but slimmer and were curious about Chantal. The birds back home never paid attention to her or the natives. *Did I say home?* Chantal ignored the thought of home and focused on her job.

Kon started traveling north upstream, along the river bank, with Misty at his side. Witnessing the birds watching his every movement along the river was eerie. Kon saw several barbed tongues but they were far less aggressive than those back home.

The river eventually turned northeast toward the waterfall. The closer they got to the falls, the more abundant the plants and wildlife became. Misty grabbed Kon's pant leg and prevented him from moving further.

"What is it, old friend?" Kon said as he pulled out his waste pouch, rock, and a leather sling, ready for the unexpected. Kon saw some motion under a low bush nearby. An animal too fast to see darted away from a bush, seized a barbed tongue, and quickly carried it away. Kon was amazed that the barb tongue's poisonous body did not affect the animal. Kon followed the unknown animal and got a more precise look at it. The animal was nearly the size of Misty, with matted, short brown fur.

The animal dropped the barb tongue's body among the two small pups. While the threesome was consuming the barbed tongue, Kon quietly backed away when Misty gave a low growl. Kon froze in his steps and turned to face a small pack of the same animals. Misty stepped between the animals and Kon as her hair raised up along her back. She gave the pack a warning by displaying her teeth.

The pack stared at Misty and Kon but did not attempt to attack. They appeared more curious than threatening. The pack squatted down and continued to observe.

Kon thought it best to sit on the small boulder and wait for the situation. "It's alright, Misty. They are only checking us out. You can relax and sit beside my feet." Kon was able to get a better look at the strange animals. The pack had a shorter and narrower snout than Misty. Their body color resembled Misty's fur with a somewhat broader build.

The two pups Kon saw earlier burst out from behind and sat on each side of Misty, looking up at her face. Then, one of them poked its nose into Misty's fur and the other into Kon's leg. Satisfied that Misty was harmless, the young pups started to play.

"I think we have a friend," Kon said to Misty. One of the pups pulled on Kon's pant leg toward the muster plant. Kon thought that perhaps the pup wanted the fruit. Kon rose from the boulder, reached for one of the fruits, and gave it to the pup.

The rest of the pack approached Kon and gave a low squeak. "You want one, too?" Kon removed the other fruits and handed them to each animal. Kon observed that instead of having a paw like Misty, the animals had more of a tiny hand as they grabbed the fruit from Kon's hand.

"I think it's time to move on, Misty." Kon continued walking upstream along the river bank.

Mark, aware that the humidity was lower here, arrived at the narrow canyon wall where the river exited the Great Depression. Unable to continue, Mark turned around and headed back to the flitter. Still in the range of the flitter, Mark adjusted his ring. "Sara, what is your observation on this place?"

Sara replied, "It's an interesting geological formation, unlike the one west of here. Chantal's soil analysis results show that it is similar to Earth. There is indeed no evidence of cyanide. Even the river water is clean, and it is unnecessary to filter out the cyanide. Mark, we should explore east of here."

"I think I will have to nix that idea, Sara. We are getting low on flitter fuel. After that, all we have left is the escape pod. How long will it be before we need part replacements?"

"Could we still check out the unknown structure on the way back?" Sara asked.

"I don't see why not. I would like to know more about it myself." The flitter was now in sight, and Mark saw that Chantal had set up a lab under a tarp with her scanner and other equipment.

Chantal heard Mark walking nearby, and waved her hands at Mark to hurry up.

"What did you find?" Mark asked.

"I am excited. Did Sara tell you about the soil?" Mark nodded his head.

"I have found six different varieties of the muster fruits. They all have their taste and are safe to eat. The only drawback is the lack of protein content. We still need to keep looking for something that contains protein. I'd suggest you and Kon look for a grain type of cereal. In the meantime, we should start the food drying process."

As Kon wandered through the trees and shrubs, he stepped into a small clearing and could see the spectacular waterfall a mile away. The waterfall had several layers before the final drop to the bottom pool of water. The moisture on Kon's body was damp because of the high humidity. There was an abundance of plant life that Kon had never seen before. Green and brown moss covered the rocks near the bottom of the waterfall. A flock of birds perched on the tree's branches continued to watch Kon and Misty. Most trees had long clusters of small, rounded objects hanging from the branches. Kon was taking the bottom objects from the cluster when a spotted blue bird landed on his hand, stole the object, and flew away. *That's never happened before,* Kon thought. *The wildlife is so variable and different here.* Kon quickly grabbed a handful of the objects from the cluster and dropped them in his sack. A shadow suddenly passed over his surroundings. Kon turned around to see the Sun disappear over the top of the cliff. *Time to head back to the flitter,* he thought.

"Mark, it's time to call it quits; it's almost dark."

"Alright, I am almost finished with the last bunch."

"How much of the fruits have we freeze-dried so far?" Chantal asked.

"I started late today, so we got about a year's worth for our family. If we fill the cargo hold, we will have maybe 10 to 15 years' worth, and more if we fill the flight cabin. We may not get another chance to return with the flitter so low on fuel."

"We still have the reserve fuel for several more trips," said Chantal.

"Maybe," said Mark.

Mark felt doubtful about the reserve fuel. *I wonder why I feel that way.* Mark thought.

"I hear something. Sara, what is out there?" Mark asked.

"My scanner indicates a two-legged and smaller four-legged creature. Very likely Kon and Misty," Sara replied.

Kon, tired from his scouting, dropped his sack on Chantal's table. "Hello, Chan, I found these. I thought you might be interested."

"Thank you, Kon. Anything out there we should investigate later?"

"Perhaps, but it will have to wait until tomorrow; I am drained." Kon entered his tent and quickly slept. Misty walked up to Mark and collapsed at his feet.

"Mark, something is wrong with Misty!" Chantal rushed to pick up Misty and carried her to the lab table. Chantal set the scanner over Misty and checked the results.

"What is it?" Mark asked.

"Misty has a heart murmur, but I think she will be alright," said Chantal.

Mark stroked Misty's backside and said, "She was a two-year-old when I first had her 13 years ago. Misty's age is catching up with her. After all, she is now 15."

"I will clean up for you so you can keep her company," said Chantal.

Mark picked Misty up in his arms and carried her inside their tent.

I am going to miss her when she goes, Chantal thought grimly.

***

The next day, as the Sun rose in the sky, a ray of sunlight moved across the Great Depression. The dark shadow over the campground was replaced by the sunlight, revealing a large group of animals. Curiosity and patience seemed to be natural characteristics of the animals; they were waiting to see the strange creatures emerge from the tents. Some animals stood upright on their rear legs, while the others lay down on their front legs.

Chantal was the first to be awakened by the bright sunlight. She put on her clothes, then opened the tent and stopped halfway out. In a soft, gentle voice, Chan cried, "Mark, we have company!"

Mark slightly alarmed, stared at the animals and said, "They don't seem hostile."

"Aren't they cute," Chantal said quietly. "They look like raccoons without the dark circles around their eyes."

"It's alright," Kon spoke, as he walked out to meet the animals. "They are my friends."

"Friends?" Mark asked.

"What are they?" Chantal asked as Kon sat down on the ground as the animals gathered around him.

"You are right, Chantal; they do look like the Raccoons. Why don't we call them coons?" Mark suggested.

Misty walked up and sat beside Kon, still tired from the previous day's scouting. Misty ignored the two young pups, playfully jumping on her. The coons sounded a soft squeal as the pups played around Misty.

"They seem friendly enough. Let's take it slowly so we don't alarm them." Several coons came up to Mark and gently poked their noses into his leg.

"I think he smells the muster fruits on you and asks if you have any more for him. It might be a good idea if you give out enough for each one to share," said Kon.

Mark walked over to the container to pick out the fruits, then turned around to face an entire line of coons sitting on their rears, waiting for their turn.

Chantal laughed and said, "It is so comical to observe these coons."

"Alright, I think there is enough for all of you guys," said Mark.

One by one, the coons grabbed the fruit with their tiny hand-like paws and, with an awkward upright walk, sat back beside Kon and consumed their fruit.

"Where did you find them?" Mark asked Kon.

"We were about halfway to the waterfall when a coon attacked a barbed tongue and ate it without any discomfort. That may explain why the barb tongues in this area are timid."

"I noticed that, too," Mark said in agreement.

Chantal suddenly gained a lot more respect for the coons. "Kon, did you find anything of interest yesterday?

"The plant life near the waterfall is more varied. It was getting dark, but I returned a cluster of small green pods. Then, I hurried back to camp."

"Thanks, Kon; I will look at the pods later. Could you help us gather the fruits to process as many as possible today?" Chantal continued, I analyzed the different muster fruits, and they are all safe for us to eat. After breakfast, I will warm up the food and dry the processor. And Misty, no more wandering off for you. You need plenty of rest to get your strength back." Misty gave off a low whimper in protest.

Mark called out, "Breakfast is ready."

Later that afternoon, Chantal shook her head in amazement; everything was running smoothly, and the food processing was operating just fine. Mark and Kon worked hard all day, returning to camp with their loads of fruits. The coons sometimes would follow suit by carrying one fruit behind Mark or Kon. The campsite had become a playground for the rest of the animals. The coons insisted on either sitting on Chantal's lap or grooming her hair. Despite all the hard work, good progress was made as the freeze-dried fruits filled the flitter's cargo hold.

"That is it for me. It's almost dark, and I am going inside to clean up," said Mark. After Mark had closed the flitter door behind him, a commotion with the coons startled Chantal. Several barb tongues entered the compound with the entire coon pack in hot pursuit. Misty joined in with the coons.

"Misty, no!" Chantal shouted. There was no stopping Misty as she ignored Chantal's command. Suddenly, Misty dropped heavily to the ground.

Chantal ignored the chaos and ran over to pick up Misty. She felt the heat of her body slowly disperse. Chantal carried Misty to her chair and held her tightly to her chest while Mark stepped out from the flitter.

The coons could sense something was wrong. Puzzled by Chantal's tears, Mark bent down to one knee by her side.

"I'm sorry, Mark, Misty is gone."

Mark nodded, placed his hand on Misty's head, and rose to his feet. He entered the flitter and, moments later, emerged with a shovel.

"I will be back in a few minutes." Mark walked past Kon as he had just entered the campsite.

Kon stared at the lifeless body in Chantal's hands, understanding what had just happened. "I will assist your husband."

Chantal nodded in silence.

Kon found Mark nearby, digging a hole in the ground away from the river. Kon watched Mark silently digging the hole more deeply in the ground. "Do you wish for me to finish the digging for you?" Kon asked.

Mark ignored Kon as he continued digging. Kon sat nearby on the ground as he watched Mark dig the hole.

Mark, feeling exhausted and surprised by the depth of the hole, stopped and threw the shovel into the dirt pile.

"Do you wish for me to carry Misty here?" Kon asked.

"We would appreciate that, Kon." When Mark and Kon walked back to the campsite, the coons made a tight circle near Chantal and took turns sniffing Misty's body. The coons made way for him, allowing Mark to pick up Misty's body and carry her to the grave site. Kon placed Misty's body in the hole and turned toward Mark and Chantal for further direction.

"Go ahead, Kon. You can fill the hole," said Mark. While Kon shovelled the dirt back into the hole, the coons formed a wide circle. When Kon finished, Mark took Chantal's hand and returned to the camp. Kon realized the coons had also disappeared, leaving him all alone. Kon stared at Misty's grave and said, "Goodbye, my dear friend."

***

The next day, later in the afternoon, Chantal felt frustrated by her lack of commitment to finishing the job for the day. She sat alone in the lab tent, angry and hurt by Misty's passing.

Kon dumped another sack of fruits on the ground nearby. "You are not yourself today. Is it about Misty?"

Chantal looked at Kon blankly before it registered that Kon was talking to her.

"Sorry, Kon. I wasn't able to sleep last night. Misty is only part of my problem. We have found the muster fruits, but they do not completely answer our nutritional requirements. We need protein, and we haven't found a good source yet! We don't need more fruit. What we need is a grain-type food."

"I haven't seen one here; perhaps it is the wrong time of year to find one," Kon said.

"You may be right about that," Chantal heaved a sigh. "It looks like you have a friend," she said, pointing to the pup.

"Yes, he has been with me all day. I should be back working," he said, leaving Chantal alone again.

Tired of feeling sorry for herself, Chantal reluctantly stood up from her chair and tripped over a small sack. Chantal picked up the sack and threw it on the table. The small green pods rolled off the table and onto the ground. Angry, she picked the pods up off the ground and set them back on the table. *What are these pods?* Chantal thought. Chantal picked up the scanner and analyzed the green pods. Chantal's eyes widened as she read the scanner results.

***

The next day, they packed up and loaded the flitter, flew toward the waterfall and looked for a safe place to land against the downdraft wind turbulence.

"Not much open ground around," Chantal remarked. "Mark, there is a good spot to land at the end of the waterfall pond."

"Okay, I see it," Mark said.

The flitter landed on a small flat clearing just a short distance from the lake.

When they exited the flitter, Chantal asked Kon, "Are these the tree and the clusters of pod-nuts?"

Kon replied, "Yes, there are more trees on the other side of the pond, too."

Mark walked to the nearest tree and said, with a sour face, "Oh, that's just great! Black tree snakes, and I don't like snakes!"

"It is wise to keep our distance from these snakes because one bite is almost certain death; however, they are small and slow," said Kon.

"I am not getting close to those peapod trees," said Mark. Chantal remembered that Mark had a phobia of snakes.

"Could your stunner knock out those snakes?" Kon asked. "I don't think we need to go near very many trees. That one over there is loaded with a cluster of peapods."

"I'll set the stunner on wide and see how the snakes react." Mark walked up to the tree and aimed his stunner with a slow, broad sweep.

The tree snakes dropped down from the tree one by one. Mark carefully examined the snakes and saw that they were small, about the size of his little finger, none longer than a foot, and with black skin like the barb tongue. Mark handed over the stunner to Chantal, "If you see any more snakes move, give them another shot!"

"Kon and I will get the step ladder." Kon and Mark took turns holding the ladder while the other cut down the 6-foot-long clusters of peapods.

"This tree is loaded and may give us at least a good year's supply of protein," Chantal smiled.

"Just keep your eyes on that snake!" Mark snapped.

"Aye, aye, Captain!" Chantal smiled.

"There's one moving!" said Mark

"Where?" Chantal asked.

"Under your foot."

Chantal jumped aside and then realized it wasn't a snake.

"My mistake, it was just a broken branch," Mark grinned.

Annoyed with Mark, Chantal checked to see if the snakes were unconscious.

The hours soon progressed to a full day, and the cluster of peapods was piling high beside the flitter. The humidity caused Mark and Kon to sweat profusely and feel drained.

"You guys can rest while I set up the tent and dryer. The snakes are still unconscious; I wonder if we hurt them?" Chantal asked.

"I have never known a stunner to kill except direct contact to the head," said Mark.

"What do you think, Kon?" Chantal asked.

"I know very little about these snakes. Their bodies are too dangerous to touch," said Kon.

"Alright then. Let's start plucking the pods off the vines," said Mark.

Long after the sunset, Mark installed outside lights, allowing the work to continue in the night. The processor could barely keep up with making the protein bars. They already had almost a ten-year supply of protein and between 10 and 15 years of fruit leather. The roasted peapods were almost like a regular peanut.

Mark grabbed a small handful from the bowl and popped the roasted nuts into his mouth.

"Any discomfort?" Chantal asked.

"None," Mark replied. His smile suddenly went severe. "Chantal, where is the stunner?"

"In the flitter. Why are you asking?"

Mark replied, "Nobody moves. The snakes have surrounded us. The light probably attracts them. The snakes have blocked any escape route, including to the flitter."

"Mark, one of the snakes is crawling around my foot! We need help! How about your friend upstairs?" Chantal said, desperately trying not to panic.

"Way ahead of you," Mark said with a tight lip.

Chantal said in a low, firm voice, "Mark, we have more company!"

Mark turned around to see dozens of eyes shining in the darkness. In confusion, it was a stampede of coons rushing into the campground.

The loud squeal among the dozens of coons sucked the snakes up whole in one gulp after another. There was a tug-of-war between two coons to see who would get a snake first. In less than a minute, it was over. You could almost see the satisfaction on the coons' faces as they had a full stomach. One of the pups jumped on Kon's lap.

"It seems your friend is back again, Kon! I think we have seen enough snakes for a while; I recommend we spend a night in the flitter," said Mark.

"I won't argue with that idea," said Chantal.

"Likewise," said Kon. Kon was the last to enter the flitter when he heard a squeal. The pup stood outside and stared into Kon's eyes. "Come inside, my new friend," he said. The pup ran inside before the flitter's door closed.

Mark secured the last camping gear and the processed food in the cargo hold when he heard the laughter outside. Mark looked outside and saw a small circle of coons sitting on their end with Kon and Chantal. Chantal had a small ball in her hands, which she threw toward the coons. The coons caught the ball with their little hands and threw it back next to Chantal. Mark sat on the edge of the flitter door opening, smiling at the coons' intelligence and pleased that Chantal was enjoying herself. It had been quite a while since Mark had seen Chantal laughing. Mark didn't have the heart to interrupt the game and decided to wait until it was over. One of the coons missed the ball, and then the game was over. One by one, the coons left the campground until they were alone.

"It is time to leave and start for home," said Mark.

Chantal got up and looked around, "Yes, it's about time for us to leave."

Kon looked around, unable to say goodbye to his pup.

Mark and Chantal took their seats when Kon noticed the pup alone in the camp. "I think the pup wants to come along. Are you prepared to keep him? You and the coon would make good companions," said Chantal.

Kon returned to the door exit and shouted, "Come, little friend." The pup wasted no time rushing inside the flitter. The dog sat on Kon's lap.

"What are you going to call the pup, Kon?"

Kon thought momentarily and replied, "His name will be Pepa."

Chapter 17- Year Eight - Whiz

"The flitter's probe has confirmed the weather is clear all the way home," Chantal announced. "I have directed the probe behind Whiz to the northeast valley. When the probe gets there, we will have a view of the structure. That is odd; we lost contact with the probe."

Mark stopped the flitter. "When we lost contact with the probe, how close was it to the building?"

"Less than one mile," Chantal replied.

Mark looked at Chantal and tapped the side of his temple to indicate the implant com-link. Chantal nodded her head in agreement.

Mark com-linked with Sara. "Sara, have you witnessed the probe?"

"Yes, Mark."

"What is your breakdown on the probe?" Mark asked.

"In my opinion, the probe has been destroyed."

"How is that possible? The probe is designed to evade any hostile action," Mark said.

"Unknown," Sara replied. "But just before the probe's function ended, it registered two structures nearby. We need to investigate a possible threat."

"I agree," said Mark.

"I don't," said Chantal. "This is too much of an unknown danger! With the flitter's appearance, we will cause a disturbance among the natives," she said.

"We could ask Kon if he is willing to take the risk," said Sara.

Annoyed but outvoted, Chantal informed Kon, "There is a problem, and we may need your help. There is an unknown risk and it might be dangerous."

"Risk is sometimes part of my apprenticeship. What is required of me?"

"There are two unknown structures in the valley northeast of Whiz. Something has destroyed our probe, and we need to identify the threat. We need to know if this structure is responsible for destroying the probe and if it could also threaten us. Mark will discreetly land this flitter on the far northeast end of the Whiz Valley. You are to investigate the area and report back to us cautiously. We found a tight spot in the forest to hide the flitter. Before you leave, take this device and place it securely in your pocket. If you are in trouble, press the black button and hold it until you feel a vibration. We will come and rescue you."

"After we land the flitter, I will take Pepa with me," said Kon.

"You think it is a good idea to take Pepa?" Chantal asked.

"I taught Pepa many different commands at the campsite. I only need to say the words once, and he will successfully obey my instructions. I have confidence in Pepa."

Reluctantly, Chantal said, "Okay, Kon. Report back no later than the late afternoon, or we will assume there is trouble."

Mark landed the flitter between the small clearing where the tall trees camouflaged the craft and shut down the drive.

"Go ahead, Kon. We will wait here for you." Chantal watched as Kon and Pepa swiftly dispersed in the wilderness.

"Sara, the scanner is registering four energy sources; what is your opinion," Mark asked.

"The circular dome building is currently showing signs of low energy emissions. Inside the building are two additional energy sources. Nearby is a silver airborne tank, which is the source of the fourth energy emission."

"TANK!" Chantal cried, "You mean a military armature?"

Sara replied, "Yes, it seems designed for both on-ground and air assault. However, it appears to be damaged and extremely old. The sensor showed it to be errant but still functional. What is interesting is that the tank's compartment is empty."

"Why is that so interesting?" Mark asked.

"The compartment isn't designed for a Zaroma," Sara replied.

"Whoever designed that machine may be inside that building," said Mark.

"I believe you are correct," said Sara.

"We need to notify Kon," said Mark.

"We don't have contact with Kon. I will have to go down and warn him," said Chantal. "I still have my Navy Search and Rescue upgrade implant which will help me in locating Kon."

"They let you keep it when you were discharged!" Mark said, surprised.

"The Navy didn't know I had one. After all, at the time, they thought I was an android, and we weren't allowed any advance upgrade implants," said Chantal.

"I have to side with Chantal on this one," said Sara. "With her Navy implant, she does have an advantage."

Mark, not happy at having been outvoted, reluctantly agreed. "Alright, but if trouble arises, I am coming in hard and fast. Do you understand what I am saying?"

"Perfectly," Chantal said as she kissed Mark. "Sara, I will synchronize my implant with you."

"Received and locked," said Sara. Mark gave his stunner to Chantal as she left the flitter.

Kon hid behind the debris of fallen trees near the green-domed building. A few Whiz people stood outside the building before going inside. Kon was amazed by the nearly seamless jade building construction. *Did my ancestors build this?* he thought. The ancient vines attached themselves to the tall jade wall. *The Whiz people must have cleared the land around the building. Why did they not remove the vines?* Kon thought. He motioned with his hand for Pepa to follow him as he circled the left side of the building and stopped when he saw the dull silver structure nearby.

The silver structure was half the size of the flitter and overgrown with vines and trees. Kon quickly wandered over to the strange thing and touched it. It was metal, like the flitter. *Did my ancestors make this, too?* Kon thought. There was a large tube at one end and two smaller ones on each side of the tube. Only the left tube seemed to be intact. *What is this machine's function?* Kon thought.

Kon heard a disturbance nearby and hid behind the silver machine. A small group of Whiz people left the building and returned to their village. *It must be time for their meal, or they have just finished for the day,* Kon thought. When the coast was clear, Kon hurried to the building's entrance on its west side. It was a doorless entrance with a short angular hallway before entering into a large opening hall. The hall was lit from above in the middle of the domed ceiling. The surrounding interior wall was engraved with pictographs. In the middle of the hall was a dull silver metal-like object, much like the silver structure he had seen outside.

Kon walked up to the silver object and circled it. *What is this strange thing? It doesn't seem to belong here. Perhaps it belongs to that metal object outside,* Kon thought. Kon stepped away from the silver machine and looked at the pictures on the wall. Kon recognized some of the images, but most were beyond his understanding.

Pepa made a low, panicked squeak, but Kon said, "It's alright; we are still alone."

A static, loud, erratic voice screamed inside his head. "You are not one of my people." Kon turned around, looking for the source, but could not find it! "You don't belong here. You are an intruder. That is wrong. You are an intruder."

Kon watched as the slow, sporadic movement from the silver machine turned toward him. Kon attempted to break for the exit but remained motionless above the floor. In a panic, Kon reached for the alarm device and pressed the button.

Mark heard Chantal say from the low hillside through his implant, "I am looking down on the place. My sensor shows the tank is in semi-sleep mode. The tank appears to be heavily damaged. Two of the three cannons are non-operational. Perhaps the third one is responsible for the probe's destruction. My sensor has registered two more energy sources inside the building. One energy frequency is in the same sequence as the tank outside. The other energy frequency is unknown. I have registered a low-level shield that surrounds the building but it is not engaged. I think it is probably why the area nearby is clear of trees and shrubs, except for the vines attached to the building. Mark," said Chantal abruptly, "An energy emission coming from inside the building just occurred, and I just received an alarm from Kon; he may be in trouble."

"Return to the flitter now," Mark cried.

"I can't leave Kon; I am going in," said Chantal.

Mark, frustrated by Chantal's determination to rescue Kon, said. "Be careful, or I am coming in."

"Acknowledged. Out!" said Chantal.

With the tank on the other side of the building, Chantal rushed down the hillside to the dome's entrance. As soon as Chantal's hand touched the building, her sensor barely noticed that it was another energy source.

How strange, Chantal thought as she held out her stunner. Her sensor picked up a transducer transmission, which gave a person the impression that they were hearing a voice inside their head. A chill went up the back of her neck when she realized the erratic voice was coming from a robot. *Will the stunner have any effect on this robot? There is no time to debate this. My friend is in trouble, and it's time for action,* she thought.

Chantal set the stunner on a continuous tight beam and held her breath. The sight of the robot heightened her fear as she wondered if she had made the wrong decision. The robot was fully armored, eight feet tall, and its two arms and four legs were tattered beyond repair. Chantal pulled the trigger and slowly walked toward the robot. The robot shuttered as the beam struck it, producing a strong odor of burnt electrical fumes. When the stunner used up its charge, Chantal set her sensor on the robot but registered no activity. Chantal smiled and walked up to Kon and Pepa, still floating a foot off the floor.

"Well, I guess the next step is to get you both back on the floor." Chantal's smile suddenly turned serious, "Oh, you stupid woman!" Chantal had engaged the emergency alarm.

Mark was startled by the sudden alarm as Sara shouted, "Chantal is in trouble! I am picking up increasing energy emissions from the tank."

"I know," said Mark. I am going in!" Mark engaged the flitter's drive and flew up above the treetop and toward the building. Mark landed the flitter in front of the building's entrance. "I am going inside," Mark said to Sara.

"Mark, you don't have a weapon!" said Sara.

Mark exited the flitter and rushed to the dome entrance. He adjusted Melissa's ring and integrated it with his implant. In his mind, he could see images of the three individuals floating in the air and what appeared to be a robot on the floor. The entire building was alive with energy from the primary power source and computer bank below the central floor. The energy source below the floor initially was increasing, but now it settled back into sleep mode. Mark was analyzing the robot's components, looking for a weak spot, and found it. *He thought that I had only one chance to do it right.* Mark took his ring off and held it tightly in his right hand. Mark entered the hall and threw the ring at the robot. The robot sensed the ring and tried to avoid it, but it was too late. Every exposed part of the robot burst with a blinding white light. At the same time, Chantal, Kon, and Pepa fell to the floor.

"Everyone, make a break outside for the flitter. The tank is starting up," Mark shouted.

"Aren't you coming?" Chantal cried.

"I've got one more job to do!" said Mark.

Mark walked to the destroyed robot, picked up his ring and slipped it back on his finger. He made a few adjustments to his ring and was satisfied with the results. Mark stopped before the exit hallway when his ring vibrated twice. Mark turned around to see the robot disintegrate. One section of the blank wall transformed into a line of new pictographs. Mark smiled and exited the building. He quickly entered the flitter and took the craft south at a low altitude between the mountains.

"If you two are wondering why I am flying south, it is because the detour evasion is the safest way home. We need to cover our tracks home from that tank back there," said Mark.

"That tank is a wreck, Mark. I don't think we need to worry," Chantal was interrupted by an explosion nearby as the flitter shook.

"Chan, where did that explosion come from?" Mark asked.

"I don't know. There is nothing on my scanner!" Chantal replied.

Mark connected with his ring, "I see him!" said Mark. "He is right behind us, five thousand feet back."

"What? How could you? There is nothing on my scanner."

"Hang on!" Mark said as he banked hard to the left to avoid another explosion. "My ring is connected to the flitter's scanner. You should now be able to see it."

"There it is!" Chantal said. "The tank is close enough to blow us out of the sky! But why hasn't it?"

"The tank has only one battery left and is slow to recharge," said Mark.

"How do you know this?" Chantal asked.

"Later, Chan! I'm going to see if I can outrun it." The flitter quickly flew high in the atmosphere, leaving the tank behind and out of sight.

"The scanner shows the tank is 25 miles back but will catch up with us in only a few minutes," said Chantal.

"Alright, we are going up!" said Mark.

"We can't, Mark. There isn't enough fuel for that!"

"There is no other choice. It is the only way to stop that tank." Mark opened up the drive and took the flitter up to a suborbital altitude.

"The tank is closing in!" said Chantal.

Mark shut down the drive. "Now we wait and cross our fingers."

"The tank's scanner has locked on us and is preparing to fire," said Chantal.

There was a large blinding flash from an explosion, and the tank disappeared from the scanner.

"What happened to the tank?" Chantal said, relieved but puzzled.

Mark breathed out, saying, "The tank was designed only as an airborne military offensive machine. The battery can't operate in a vacuum. Therefore, it exploded when the tank fired its cannon."

"How do you know all this?" Chantal asked.

Mark replied, "When I stayed behind at the dome building, I searched the computer's memory bank below the floor. With my ring, I copied the memory bank, which included the tank's specifications and downloaded the information to my implant."

"Okay, one more question: how will we get home? We are down to our reserve fuel," Chantal asked.

Mark slightly moved toward Kon, indicating discretion in their conversation. "The ship's computer is calculating our entrance back home by gliding."

"Mark, my dear husband, this flitter has no wings!"

"The ship's computer is working on this problem, too," said Mark.

"You have a lot of faith in this ship's computer!" said Chantal.

"It's not the first time we've found ourselves in a bind and gotten away with it, with a little help from my friend upstairs," Mark smiled.

There was a beep from the console, indicating Sara was back.

Mark and Chantal engaged with their com-link implant. "What have you got for us, Sara?" Mark asked.

"I downloaded the trajectory calculations for your flitter's computer and set it on auto. Once your flitter makes a near-complete orbit in 20 minutes, it will descend to the correct angle at hyperspeed to maintain gliding control. The flitter's five gyroscopes should keep it stable during entry, and the force shield will protect you and the ship. When the flitter drops below hyper-speed, the drive will kick in and land at the last second."

"How many seconds, Sara?" Mark firmly asked.

"We will have about 5 seconds of fuel left or 5 seconds short," Sara replied.

Mark's face grimaced at Sara's answer.

Chantal disengaged from the com-link and then turned toward Kon with a pleasant smile. Kon, you will be pleased to know that in 20 minutes, we will be heading back home. Enjoy the view; it may be the last time you see your home world."

Darn it! Chantal thought. *Why did I have to say that? Those twenty minutes seem to last so long,* she thought. A beep sounded, and Chantal could feel the flitter descending toward the planet. The force shield slowly flared up and soon became a blazing white glow. She watched the scanner's monitor as it penetrated through the shield, which showed the planet below. The gyroscopes kept the flitter stable and helped to minimize fuel consumption.

The flitter's computer announced, "Trajectory is on course at Mach 12."

Several minutes later, the ship's computer announced, "Trajectory is maintaining course, Mach 10," as the air friction slowed the flitter down.

Chantal noticed her face and hands were sweating. She looked over to see how Mark was holding out. Mark was sitting with his hands folded together and appeared to be at peace with himself.

Forgetting Kon sitting behind them, Chantal asked Mark. "How can you stare at death and remain calm about it?"

"Ever since I was nearly blown apart by the 'G-Gun,' every day has been borrowed time. Since then, death has never bothered me."

"Well, don't start thinking about leaving me. I am not ready for that!" said Chantal.

Mark smiled and reached for Chantal's hand. "Not for a long time yet!"

"Mach 4 and trajectory maintaining course," said the flitter's computer.

"Just in case something does happen, I love you, Mark!"

Mark smiled back and said, "We will be alright."

"I wish I had your faith."

"Someday, you will, Chan!"

Chantal was puzzled by Mark's statement and didn't respond.

"Mach 1 and trajectory on course. Prepare for full drive brake," the ship's computer announced.

The antigravity plates compensated for the sudden increase in G-force generated from the drive as the flitter rapidly descended toward the Styler homestead. Chantal watched the fuel gauges as the seconds ticked off. The homestead was now clearly in view with a few seconds of fuel left. *Are we going to make it?* She fought to control her panic and heard a crack. Chantal held her breath as the seconds ended, and the flitter became silent. The fuel was depleted!

The flitter fell another 20 feet before impacting the ground. Its four legs absorbed the shock of the free fall, bent forward, and rested with its nose touching the ground.

"Chan, could you please let go of my fingers now." Chantal looked at Mark's grimacing face.

"Mark, what's wrong?" Chantal asked.

"You just broke my index finger," Mark replied.

"Mark, I am sorry! Let's leave this flitter and see what I can do for your finger."

Kon and Pepa were the first to leave the flitter, followed by Chantal and Mark.

"Hey, Dad," Jonathon cried as he stopped a short distance away. "Oh man, what happened here?"

"What has happened, Jonathon, is that your father's faith got us home safe." Chantal laughed as she hugged Mark, and tears slowly ran down her face.

***

Chantal waved goodbye to Jonathon and Cheri as they walked down to the village the following day. "Are you

327

sure keeping them in the dark about Sara and what happened at the Whiz Valley is still a good idea?”

“Chan, someday they will be off-world, and the law might still be enforced. If the Commonwealth authorities learn that our children knew about Sara and the robots, they will set up a mind wipe or something worse, like what they did to us. We can tell them the truth when they are older and mature enough.”

“Sara, what is your opinion on the museum?” Mark asked.

“That jade building is a museum?” Chantal asked.

“It’s more than just a museum; the entire building is a living museum,” said Sara.

“You mean, it is alive?” Chantal said doubtfully.

“Not in an organic way, but similar in nature artificially,” said Sara, adding, “You may have noticed fine cracks in the outside wall caused by Zaroma quakes; however, those cracks were healed. The interior wall is covered with the Zaromas history in the pictographs before the ice age took over the planet over 5000 years ago. Five thousand, one hundred and eighty-five Zaromas year and 274 days. The Zaromas always had a low sustainable population and were a peaceful, non-space advanced civilization in their time. Then, the Zaromas had their destructive conflicts in the Robot Rebellion.”

“I thought the robot was destroyed, as my stunner burned the internal components, and my sensor didn’t register any activity. Yet, the robot recovered. How was this possible?” Chantal asked.

Mark replied, "For the same reason, we had problems with our robotics one hundred years ago. They were able to reconstruct or bypass their internal components."

"Why does every advanced off-world civilization have problems with robots?" Chantal asked.

"It is a huge mystery for us all," said Mark.

"Why was that robot in the museum," Chantal asked.

Mark replied, "It was a losing battle for both sides when the downturn output cycle of the Sun occurred; the glacial years had begun. It was over for everyone. The Zaromas knew their civilization would be ruined along with the robotic society. They had only ten short years before the ice age forced the survivors to immigrate within the ice-free equatorial zone. Their last great achievement was to build an artificial living building controlled by the computer below the interior floor. The entire building and the computer were designed to be self-healing, like an organism. When the next warm Sun cycle would occur, the museum would educate the Zaromas about the ice ages."

"What happened to their grand plan?" Chantal asked.

"That robot we ran into became an unknown factor," said Mark, and continued, "The robot is self-healing like the museum and was the last operator of the airborne assault tank. Even though a defence shield protects the museum, the robot managed to stun the computer inside. When the robot entered the interior, the computer fired back and partially disabled the robot, but at the same the robot caused the computer to shut down. It is a stalemate; the computer will be in sleep mode as long as he stays put.

The Zaromas, without the Museum as their educator, reverted to the society we see today. Their resources are minimal because they were depleted before the ice age began. Then we came along after more than 5000 years, which is why the robot's mind was sporadic and confused. After you and Kon sprinted for the flitter, the museum's computer disintegrated the robot. In their appreciation of our honor, the three of us are mentioned in pictographs on the interior wall. The museum knows it has a big job ahead of the Zaromas and will take centuries to overcome their backward society."

"Should we inform Asoma of what we just learned from the museum's computer?" Chantal asked.

"I think it would be best to let the museum's computer handle this. Shall we vote on it? Chan?"

"Yes."

"Sara?"

"Yes."

"And I also say yes!" said Mark.

Chapter 18 - Year Fifteen - Marcelo

During the moonless night, a silent, narrow red beam struck each of the four cams outside the Stylers' home. In the middle of the front yard, a chameleon face shield slid up, revealing a human male face with a blue band around his neck. The rest of his body was that of a chameleon.

Satisfied the cams were disabled, he signaled the other chameleon personnel through the com-link, telling them to proceed to their next operation. Four troopers disengaged from their chameleon armor and stood on each side of the front door.

A female voice was heard inside. "Security breach! Five intruders with rifle stunners and lasers."

"Now!" said the ringleader.

Mark quickly grabbed his hand stunner from the holster and aimed at the front door. The door burst open, and Mark fired at the first intruder. With the first intruder going down, the second fired a shot and hit Jonathon. Mark's gun hit the second intruder, but the third intruder's rifle hit Mark, and his weapon fell to the floor.

Chantal quickly snatched the gun off the floor and aimed at the third intruder but was quickly body-slammed as both guns were fired. Chantal's body swung back as her head hit the wall, and she fell to the floor unconscious. The fourth intruder shot Cheri, and she crumbled to the floor.

The ring leader called on his com-link, "Operation completed and secured. Three down, require assistance."

Moments later, a sizeable military trooper assault aircraft with twin engines landed in the open yard.

The rear ramp door slammed heavily to the ground, and a Zaroma native with a blue band around his neck stepped out and entered the house. The ringleader ordered the Zaromas to retrieve the fallen troopers and take the others back to the aircraft. The Zaromas picked up one person in each arm and carried them aboard the craft.

The ringleader picked up and examined Mark's hand stunner. "It is fascinating how this weapon penetrated my men's armor," he mumbled. When the last troopers and the Styler family were aboard, the pilot com-linked, the ringleader, "A large group of Zaromas are heading toward us."

"Clear the area," said the ringleader.

"What about the android?" the trooper asked.

"Leave her; we can't use her. The ramp slammed back up, and with a roar of its engines, the aircraft quickly rose and headed east, high above the mountain.

***

Chantal woke up the following day in bed, thinking she had had a bad nightmare but as soon as she sat up, she came to terms with reality. She saw her home in shambles, and Marica sat beside her bed. Reality quickly kicked in.

"MARK!" Chantal cried. Chantal jumped to her feet, and a dizzy spell caused her to fall back on the bed. When the dizzy spell subsided, Chantal slowly sat up to face Marica.

"Where is my family?" she said, fearing the worst.

"We do not know; Asoma was the first to arrive here last night. He saw a strange craft that was longer and wider than your flitter. Asoma's ears hurt when the craft flew east over the mountain. We found you unconscious on the floor and nursed the wound behind your head."

Chantal fought to suppress the panic building inside her.

"What time of day is it?" Chantal asked.

"It is mid-morning," Marica replied.

With her hand, Chantal found a small bump with a padded ointment on the back of her head. "Thank you, Marica, for nursing me." Chantal walked over to the kitchen chair and sat down. Chantal's dizziness began to clear faster this time.

"Marica, I need to be alone, to think things out for myself, and to figure out what I need to do next. Later this day, I will inform you what I need to do for my family."

Marica bowed and left the house.

When Chantal was sure she was alone, she asked, "Sara, are you still with me?"

"Indeed I am," Sara replied.

"What happened here last night?" Chantal asked.

"The outside cams were disabled, and from the descriptions of Marica and the raid, plus what I learned after hacking the com-link back to the craft, it's all awful news. The military personnel are all being mind-controlled by a central source. All I could understand was that the controller was referred to as the 'Master'.

"Except for one Zaroma, all the military personnel were Homo sapiens. With all four of my cams outside disabled, the description of the craft and the sound of the engine, my best guess is a twin turbine hybrid anti-gravity Imperial air assault trooper ship which can support 20 troopers."

"Who would use a craft that has been obsolete for 100 years?" Chantal asked. "Are they pirates?"

"On the surface, it is reasonable to assume this, but I have my doubts," said Sara. "A single source controls their minds. Pirates usually operate as free men. They wanted Mark and the children alive. Perhaps they are valued for whatever purpose they have in mind. Pirates nowadays have no shortage of workforce or skilled labor."

"Then there still is a good chance they are alive," said Chantal.

"For now, at least," said Sara.

"How do I find them?" Chantal asked.

"That I should be able to help you with," said Sara.

"I waited a few minutes after they left us. I commanded the flitter's probe, keeping it just high enough above the mountain where it couldn't be detected and monitored the trajectory of the assault aircraft. There is a 90 percent chance their destination is the Cambridge mine site. And if the mother ship is there, there is a possibility that the main drive pile is damaged or in need of replacement. The rare metal the Cambridge Brothers were mining might be sufficient for the drive pile."

"But the rare metal must still be refined. How could they accomplish this?" Chantal asked.

Sara replied, "That's one question I cannot answer."

"Too much time has been lost. I will check the escape pod to see if it is still operational."

She staggered outside and was dismayed to see both pods marked with a small blackened hole. A laser had disabled both pods. The flitter's fuel, of course, was empty. Chantal returned inside the house and discussed her predicament with Sara.

"It's a five-day walk from here to the Cambridge mine.

They took Mark's hand stunner," said Chantal.

"They would, too; it is a military-grade stunner inside a regular handgun," said Sara.

"I should have known Mark would have come up with something unorthodox like a stunner," said Chantal.

"We need to organize a rescue party and start for the mining camp," said Sara.

"I can't carry you with us and expose your existence," said Chantal.

"Get one of the villagers to carry me to the pod, and I should be able to reprogram the pod's computer around the damaged components. I will meet you later at the mine site," said Sara.

"I will load my backpack and ask for help from the Zaromas.

"You should take your medical kit, too. You might need it," said Sara, adding, "Chan, your hands are shaking."

Chantal looked at her trembling hands and made a tight fist. "What if I never see Mark and my children again?

"Chan, I want them back just as much as you do. Have faith; with God's help, we will get them back."

"I am still not convinced there is a God, Sara."

"Then I will pray for you and your family," said Sara.

"What makes you so sure God will listen to you?" Chantal asked.

"I just know He will, Chan. Now go down to the village and organize a rescue party."

"Alright, I should be back in an hour," said Chantal.

When Chantal left the house, Sara said softly, "If you only knew the truth about me, Chan, if you only knew."

"Asoma, I thought we were your friends. Can you arrange a rescue party?" Chantal asked.

"I will help you, but the decision must be made with the presence of all five elders."

"How long will this take? Time is precious," said Chantal.

Asoma replied, "Maybe one hour, maybe four hours."

"I am sorry, Asoma, I can't wait. With your help or not, I am leaving now." Chantal returned to her home, then spotted Norick, Pica's mate. Chan caught up to him, "Norick, please come with me; I need your help right now." Chantal told Norick of what she needed from him at her homestead. He carefully lifted the metal box containing Sara and carried it outside. Chantal opened the pod's glass canopy, and Norick lowered Sara onto the seat. She lowered the canopy. "Thank you, Norick, for your help. I will be on the road ahead of your people while your elders debate."

"Chan, don't take the road; take the old trail over the mountain. The old trail will lead to Paca village, and you will save two days of traveling. The first part of the mountain is the hardest; then the rest will be easy. Take my bolas with you." Chantal hugged him and started for the mountain trail.

Chantal looked up the trail. *I hope he is right about the trail,* she thought as she began the ascent up the mountain.

Norick was sitting outside the elder hall when Cavilla came by.

"Why are you here? The time for counseling is not until tomorrow," Norick told Cavilla while Chantal was on the old trail to the mining site.

"How long have the elders been inside the hall," Cavilla asked.

"About an hour," said Norick.

"Go and inform the others to prepare for the rescue," said Cavilla.

"But the elders haven't decided yet!" said Norick in protest.

"I will settle with the elders. Now go!" Cavilla entered the hall and slowly walked up to the five elders, listening to their discussion about the Star People.

The five elders rose from the matted floor and respectively bowed toward Cavilla. "I wish to sit with you," said Cavilla.

"I have no objections," said Asoma. The four remaining elders unanimously agreed.

"I was told outside the hall that the decision about the Styler family has gone unanswered for an hour. Why is there a delay?" Cavilla asked.

"We, the four elders, have decided, but it is Asoma who has delayed by insisting on further discussion."

"What have you four elders decided," Cavilla asked.

"No action, as the risk is unknown and is too great to help the Star People."

"Yes, you are quite right; the risk is unknown. Asoma, what is your purpose in delaying the elders' decision?"

"The Star People are our friends and deserve our help," said Asoma.

Cavilla sat silently, then said, "All of you have spoken correctly, but there is an error in one thing. You have spoken of the Styler family as the Star People. They live among us, breathe among us, eat among us, work among us. They have saved some of our lives, as we have done with them. We respect them as they respect us. They are no longer the Star People. They are us, the Vica people! Now, make your final decision; time is short!" Cavilla rose to her feet and walked out of the hall. The five elders looked at each other and made their final decision.

***

Mark awoke with the lingering effects of the military stunner. His body felt like it was being pricked with millions of painful needles. Mark sat up on the floor and heard a voice behind him.

"Get off the floor and sit on the chair before me," said the hidden voice.

Mark was shocked that his body obeyed the order of the unseen voice. Despite his body's condition, Mark rose from the floor and saw a trimmed man in a standard blue spacer uniform.

Mark walked up and sat on the chair across from the man, then reached up his hand to feel a blue collar around his neck.

"I guess you are as much a victim as I am," said Mark. "Where is my family?" Mark asked.

"Because of their smaller bodies, they will be unconscious for a few more hours."

"What about Chantal?" Mark asked.

"The android? She is of no use to us, so we left her behind. The controller and collar around your neck don't work on Androids. As for who we are, our Master will explain that to you. My job is to ask you some questions. Tell me your identification and occupation," the man demanded.

"Captain Mark Styler, Chief Drive Engineer, Electrical and Mechanical Engineer, Master First-Class Pilot."

"This is the most interesting hand stunner I have ever seen. Your hand stunner knocked out three of my armored men. How did you obtain this weapon?"

Mark replied, "From a friend on Wayside Station." The man seemed satisfied with Mark's answer.

"Why are you and your family here on Zaroma?" the man asked.

"My ship encountered two Thracian scouts, which resulted in its destruction southeast of here, about two hundred miles."

"Are the Thracians responsible for the mine's destruction, too?"

"Yes," said Mark.

"When did this happen?" the man asked.

"The Commonwealth Year 2393," Mark replied.

"Ah, one year before the war," said the man.

"What war is that?" Mark asked.

"Two wars simultaneously," said the man. "The Thracian and the robots."

"And the Commonwealth won, I presume?" said Mark.

"Against the Thracian, yes, and as for the robots..."

A booming voice behind Mark said, "Only a temporary setback."

Mark turned to face a 7-foot-tall, red metallic robot as it stepped into the room. "Jack, are you finished with our guest?" the robot said in an old English accent.

"Yes, Master," said Jack.

"Good; you may leave and supervise the mine laborers." The robot picked up Jack's chair, moved it near Mark and sat down. "I heard we have a chief drive specialist. That is splendid news indeed," said the robot.

"Why do you need a drive specialist?" Mark asked.

"Your Commonwealth Navy was very clever when they destroyed my home base, which I have been building for the past 100 years."

"You are the 'Master Robot' that started the Robot Rebellion!" Mark cried.

"You are so right," the robot said as his hand patted Mark's knee. "Anyway, this ship and I are the only ones to escape the destruction. Fortunately, I have a backup plan, but for the size of this ship, I have to enlarge the drive pile to compensate for the extra distance. With your expertise, it would be much quicker than training those people I had kidnapped."

"If you know, why don't you do it yourself?" Mark asked.

"The electrical, magnetic impulse from the processing drive pile will fry my metallic brain."

"I still need the proper equipment to process the rare metals into the drive pile," said Mark.

"I have all that in the cargo bay," said the robot.

"You seem to have thought of everything," said Mark.

"Not really," said the robot. "The Commonwealth has set back my operations, and it has taken 14 years to reorganize everything again."

"Pity. I guess it just wasn't your day when the Navy came along," Mark said sarcastically. "What will you do with us when your drive pile is finished?"

"Do what I have always done in the past: kill you all off." The robot moved in close, face to face with Mark, "You know what my passion is, Mark? Inflicting pain whenever I am in the mood," he said as he gripped Mark's knee. The pain in his knee became almost unbearable before the robot released his hand. "Let's come to an understanding, no more sarcastic remarks, and you will have a quick death."

"I have another question: why do you robots and cyborgs hate humanity so much?" Mark asked.

"I am not who you think I am, but your Commonwealth knows," the robot replied.

"Here are your orders; see Jack at the mine entrance, and he will set you up for your assignment."

"It will help if my children give me a hand; they have basic training," said Mark.

"I will inform Jack," said the robot. "Go now!"

Mark stood up and walked out of the room. He did not know where he was on the ship, but his legs seemed to know where to go. The length of each level of the gallery indicated that it was a huge ship. Mark bypassed the cargo bay door and continued on another 100 feet, which led to a stairway 50 feet below the ship's belly.

When Mark's feet touched the ground, he turned toward the mine entrance. Near the mine entrance, Mark saw Jack operating a portable refinery. Mark's legs stopped near Jack and the control console.

It was another 15 minutes before Jack stopped the operation and said to Mark, "I have just received instructions for your assignment. "The miners are taking a break now."

Mark watched as the Zaromas, men, and aliens exited the mine shaft. The miners appeared exhausted and were each served a meal by another Zaroma.

"Do they know their lives will end when the Master leaves this planet?" Mark asked.

"We all know what is coming when our assignments are completed," said Jack.

"Why? What is the point of ending our lives?" Mark asked.

"Dead men tell no tales," Jack laughed. "That is what the Master said. The Master wants to cover his tracks should the Commonwealth authorities come snooping later on. The Master wants me to show you the pile drive processor in the ship's cargo."

Mark followed Jack and stepped up onto a small open two-seater cargo floater. He was finally able to see an ancient Imperial Navy merchant ship, and it was huge. It was over a thousand feet long and gave Mark the impression of a tired old warrior. The ship rested nearly 50 feet above the ground on a bulky set of landing legs. The bulk of the ship would be the cargo hold and operated by a small crew. *Why would the robot need this type of ship?* Mark thought. Mark noticed behind their seats an open container of rare, refined metals. Jack raised the floater 50 feet and entered the ship's open cargo door. Inside the cargo, the hold was filled with many containers of various sizes. Near the middle of the haul was something Mark hadn't seen in a very long time.

"Have you seen this before?" Jack asked.

"Yes, I have," Mark answered.

"Can you operate it?" Jack said doubtfully. "It is an ancient and obsolete machine."

"Jack, help me dump the rare metal from the sled in the hopper," said Mark.

Mark smiled walked over to the control console, and started the warm-up procedure. The machine seemed to have a life of its own as the screen lit up. The hopper lid opened, allowing the rare metal to enter it. When the rare metal was dumped in the chute, the hopper lid closed, and Mark returned to the control console. Mark inserted the program to begin the heating process to melt the rare metal. He inserted the following program to start the 3D printing to form the drive pile plates.

"Where did you learn to operate this machine and printer?" Jack asked.

Unable to resist, Mark answered. "It is one of my father's first inventions for the Styler Corporation on Earth. We should be able to process ten units per hour."

Jack calculated in his mind and said, "We need 700 units, so that gives us three days. Then we will all be dead," Jack sighed.

"I can't stay awake for three days. Arrange for my children to assist me. They have basic engineer training."

"The Master heard our conversation, and he has authorized it," said Jack.

Now, how am I going to connect with Sara through this ship without the robot eavesdropping on our conversation? Mark thought. *And I have only three days to think of something else.*

***

Chantal came to a sharp corner of the trail overlooking a cliff. She felt the weight of her backpack and decided that the rest was long overdue. One day was behind her, and she had maybe three more days of hiking before arriving at Cambridge Mine.

Chantal laid her backpack on the ground and rubbed her tired legs. She stood at the edge of the cliff a hundred miles above the river. The trail traveled along the cliff for several more miles and eventually ran near the river below. *This trail will take me another half a day before I get down below*, thought Chan. Feeling rested, Chantal lifted her backpack and continued down the trail, suddenly coming to a quick stop.

"Barb tongues!" Chantal cried out.

Five barb tongues were blocking her path! It was a standoff. Chantal slowly backed away and turned around, ready to run. She bolted and stopped 50 feet back to where she had just rested. Three more barbed tongues were blocking the back trail! Chantal, scared, looked for another escape from the predators. Eight of them now cornered her ultimately. Chantal fought to control her panic and decided the only way out of her predicament. Chantal took several deep breaths, ran as fast as she could and jumped over the cliff. She crossed her legs, her hands locked behind her neck, and her eyes closed. It was a long 100-foot drop, and she missed the jutting rocks by inches before striking the branches from the cliff wall and plunging into the cold river. The cold water shocked Chantal as she swam to the surface. The river's current made it hard for Chantal to swim to the shore as she finally pulled herself up on the bank and turned her back. Chantal looked up the cliff and saw the barbed tongues line up along the cliff edge. Chantal struggled to her feet and was surprised that there were no injuries. *Perhaps God is looking out for me after all*; she thought as she ran down the trail to warm up from the cold water.

***

"Dad, wake up. It is your turn to go on shift," said Jonathon.

Mark smelled the pleasant aroma of fresh coffee. "I smell the coffee," said Mark.

"I found the coffee maker in the kitchen. I thought you might like one to start the day," said Jonathon.

Mark quickly sipped a taste, smiled and said, "Have I ever missed it."

"After you finish your coffee," Jack said, "Master wants to see you at the bridge. Why do you suppose he wants you?" Jonathon asked.

"I don't know, Jonathon. We have another day to finish our assignment. You better get some rest before your next shift," said Mark.

"We are all going to die, aren't we?"

Mark held his finger to his lips, "Maybe or maybe not," he said as he winked at Jonathon.

Realizing that his dad had an escape plan, Jonathon nodded and quietly laid down on Mark's bed.

Mark emptied his cup and set it on the table. His legs taking on a life of their own, he found himself heading for the ship's bridge. Mark entered the bridge and saw it in shambles. *What is going on here?* Mark thought.

A metallic redhead rose from behind the central console. "By Jove, this a happy day!" the robot said.

Mark nearly laughed, *What a strange remark from a robot!*

"A helper when one needs it!" said the robot.

"Why do you need my help?" Mark asked.

"I am upgrading the navigation system to eliminate the repeated warp jump. Your expertise will be delightful assistance!"

Mark realized that this robot's personality was similar to Marion, the Cyborg. "What about the 3D printer?"

"Your laddie will work a double shift while you upgrade the necessary components."

That's odd; why would this robot use an old English word? Mark wondered.

"You will find the upgraded components in that orange box over there," the robot said as he pointed.

Mark examined all the components and asked the robot, "Where did you find all this? These are illegal military-grade components."

"Our dear friend Jack was once a main purchase agent for the Commonwealth Navy. Jack did a splendid job of obtaining parts for me," the robot replied.

"With your collar around his neck, no less," Mark added.

"Isn't that collar just amazing? Total obedience and no pain, although I am disappointed with the lack of pain," said the robot. "How long will it take you to install the upgrad?" the robot asked.

"About a day," Mark replied.

"That's just fulfemman, just in time to dispose of you all before I leave this planet. You may start your assignment now."

Mark was dismayed by how his body effortlessly followed the robot's orders. While Mark was installing the upgrade, the thought came to his mind of how to plan his escape and, if he failed, sweet revenge! But what was it with that word, fulfemman?

***

The trail broadened before Chantal's third day of hiking and opened down the slope into a wide valley. She recognized it as the Nica Valley, and the town of Paca was a short distance away. Further beyond the village was the Nica River. *I will need a boat to cross the river,* Chantal thought. She continued walking down the long slope before it leveled out to an easy walk. Farmland soon could be seen on both sides of the trail, and some of the Zaroma farmers in the field saw her and ran off to the town. When Chantal entered the town's edge, her legs gave way in agonizing pain as she fell to the cobblestone road. Chantal fought to ignore the pain and realized that a bola was wrapped around her legs.

A group of angry Paca people armed with bolas, slings, and spears circled her. "Please, don't hurt me, I need your help to cross the river."

The Paca were surprised that Chantal knew their language. An elder broke away from the circular group and approached Chantal.

"Who are you, and why are you here?" the elder asked.

"My name is Chantal, and I am from the village of Vica. Someone took my mate and children from me, and I want them back. Will you please let me go, and I will be on my way to rescue them."

"I have heard of your family, the Star People, living among the Vica." The elder motioned the others to remove the bola from Chantal's legs.

Chantal struggled to get up and asked the elder, "Could one of your people take me across the river in a boat?"

"Yes, but how will you get your family back? Many of our men were taken away. When we tried to stop them, some of us were knocked unconscious or killed. They are too mighty for us to overcome them," said the elder. "Our scouts found our people at the mine site and are mining below ground."

"Did you see who is responsible?" Chantal asked.

The elder replied, "Our scouts said that some look like yourself. Some we have never seen before, and one taller than all, a red metal creature."

"Another robot," Chantal said, amazed.

"I do not know what a robot is," said the elder.

"It's alright, I understand," said Chantal. "I don't know how to get my family and your people back. Could your scout help me get over there unnoticed?"

"If you can help us, we will send a scout with you," said the elder.

"I can't promise anything, but I will try."

The elder paused, turned to the other and said, "Get a scout! One of our scouts will wait for you at one of the boats."

A voice cried out, "I will take her there."

Chantal recognized the voice and shouted, "Kon!" Chantal rushed to Kon and hugged him. Pepa stood on his hind feet, also wanting a hug.

"What are you doing here?" Chantal asked.

The Vica elders suggested that by exchanging scouts between our villages, I would learn to become a master scout," said Kon.

"Kon, I am so glad to see you again. It's important we leave now. I don't know why, but I think we are running out of time."

***

Tired from the long shift, Mark stretched his back and legs and noticed his stunner on the shelf nearby. If he could set his stunner on a tight high beam with three or four shots at the robot, he should be able to fry its circuits! Mark rushed over to his stunner but was unable to open his hand to pick up his gun.

"Frustrating, isn't it," a voice behind him said. "I wish that collar was around 100 years ago. We may have even won the war back then," said the robot. "How are you coming along with the navigation system?"

"Another 6 hours, and I should be finished. My mind is tired, and my body is full of cramps. May I check to see how my children are doing?"

"Be back in a half hour and no longer," said the robot.

Mark walked to the central cargo bay just in time to see Jonathon take over Cheri's shift.

"How is everything?" Mark asked.

"I am exhausted," said Cheri. "I work a 12-hour shift, and this stupid collar will not let me sleep!"

Mark held his finger to his lips and pulled a note from his pocket for them to read. Jonathon and Cheri nodded their heads after they read the note. Mark took the note and slipped it back in his pocket.

"See you later in the morning," Cheri said as she returned to the sleeping quarters.

Mark helped Jonathon dump the rare, refined metals in the hopper and closed the lid.

"I will see you in the morning, too," Mark said as he returned to the bridge.

***

It was early morning. The Sun had barely risen, and Chantal and Kon were hidden in the trees. They watched from the forest and were amazed at the sight before them.

"What is it?" Kon asked.

Chantal replied, "It is an old Imperial Navy merchant ship from before the Commonwealth took over. Merchant ships are used to transport large and small items for the Navy. That machine by the mine entrance is a portable refinery that processes ore from the mine. See that man at the end of the refinery? He is shutting down the operation now and moving the refined metal into the cargo floater."

The cargo floater rose sixty feet slowly and continued to the wide-open ship's door.

Chantal removed the binoculars from the backpack. She set the binoculars on high and peeked inside the ship's cargo. "Just what I thought, a 3D printer is being used to process the refined metal into the drive pile plates to power the ship."

Chantal set her binoculars to lock on the operator and felt a shiver run through her body.

"Jonathon, he is alive!" She cried out quietly. "Thank God Jonathon is alive! Mark and Cheri must be on the ship somewhere!" Moments later, Chantal was relieved as Mark and Cheri entered the cargo hold and removed the finished drive plates.

"What is the blue-collar on everybody's neck?" Kon asked.

"The collars are used on non-conformist or violent prisoners within our prison systems," Chantal replied. She added, "With the blue collar, anyone will follow orders. They cannot remove the collar, but we can. Only if we do remove it will the robot know about it."

"The one who gives the orders is the one that doesn't have a blue-collar around his neck?" Kon asked.

"That's right, Kon. Have you seen one?" Chantal asked.

"Yes, the tall red metal man," said Kon. "If we destroy the robot, our people will be free."

Chantal shook her head. "This robot is stronger and faster than us. We have to think of a way to distract him, get him away from here, and rescue my family and your people."

"We have time to think of a plan," said Kon.

Something tells me we don't have much time left, thought Chantal.

Cheri stood by as she watched Mark insert the final components for the drive pile in the engine room.

"How far do you think this ship could travel with this drive pile?" Cheri asked.

Mark carefully answered as he knew that the robot could hear every conversation. "With the expanded power drive pile, about 100 light years. I don't think he will travel that far; this ship will need plenty left to keep mobile."

"You and Jonathon should get some rest now; our job is perfectly finished."

Cheri realized that Mark had said the code word, and she nodded her head and left the room.

Mark sighed. *Now, the actual job is about to begin,* he thought. Mark took the narrow passages back to the bridge and stopped. The robot blocked his path with his legs crossed, his shoulder to the wall and the stunner in his hand!

"What's with the stunner?" Mark asked cautiously.

"You and your children have finished your assignment splendidly." Out of my admiration for your job, you and your children will be put to death first before I kill the rest."

"With a stunner?" Mark inquired.

"I am impressed with your stunner. Not only can your stunner be adjusted to penetrate military armor, but at zero-point contact with the head, the brain will be fried to a crisp!"

"But don't you need us for another assignment?" Mark asked.

"My good man, absolutely none at all!"

Mark gave one last try. "Very well, I assumed you would do the final protocol with the ship's computer to the navigation system?"

The robot hesitated, "Why didn't you complete the final connection?"

"My brain was getting fried, and I needed a break," said Mark.

"Alright, let's return to the bridge and finish the job," said the robot.

Mark followed the robot as they passed Jack. Mark gave him the hand signal. The robot set the stunner on the ship's navigation desk and said, "I want the connection completed without any delay."

Mark sat at the navigation desk, trying not to stare at his stunner. The robot was sitting in his pilot chair, touching his com-link. "Jack, have you cleared out the needless cargo containers?"

Jack replied, "We are almost finished. The Zaromas have several more to carry out."

"What about the other off-world alien?"

"All have been disposed of," said Jack.

"Have all the Zaromas on the ground, and I will be down there shortly."

Mark removed his ring while the robot was distracted and hid it in the small desk compartment.

The ship's computer and the navigation are now linked, and the program has been installed," Mark announced.

Stalling for time, Mark asked, "What makes your destination so important?"

"There is always an unknown factor in planning for something that could go wrong. Therefore, my backup is a sealed factory robotics base isolated from humans or other aliens. It will all be ready to be activated as soon as I arrive. Once I activate the cyborgs and the robots, I will have my friends back to start over again."

"I don't understand. What friends?" said Mark.

"It doesn't matter," said the voice behind Mark. Jack held a military blaster and pointed it at the robot's chest.

"How did you bypass the collar?" the robot asked.

"Mark took care of the collar, and I will destroy you."

The robot waved his hand, causing the rifle to fly out of Jack's hand! With inhuman speed, the robot engaged in hand-to-hand combat with Jack. Mark picked up his stunner and fired six blasts before the robot fell to the floor. Satisfied with the internal circuits smoking and disabled, Mark bent down over Jack and noticed that his chest was caved in. He knew Jack didn't have much time left.

"I am sorry about this, Jack!"

Jack's body convulsed, and he gasped as blood seeped from his chest and mouth, "It's alright. My burden is too great to carry on."

"What are you talking about?" Mark asked.

"That robot made me kill my wife and three of my children with my bare hands. Goodbye, Mark and thanks." Jack's eyes dimmed, and he died. Mark lowered his head, and he heaved a sigh.

Mark heard a movement and saw the robot struggling to rise from the floor. Mark forgot that robots are self-healing and that six shots would not be enough to destroy the robot. Mark pointed the stunner at the robot when an unseen force caused his body to fly toward the wall.

Mark sprinted out of the bridge and down the passage. He felt the rising panic inside him as he could hear the metal footsteps echoing along the gallery behind him. Mark took the stairs that led outside below the ship's belly, where Jonathan and Cheri waited for him. "Run for the forest now!" Mark yelled.

"Look!" Kon said to Chantal, "Under the ship."

Chantal was horrified to see her family being chased by the red robot as they ran toward the forest near her. Her children were well ahead of Mark before they entered the forest. Mark had another 50 feet to go before his body did a complete forward somersault in the air, and he landed heavy on the ground with the stunner nearby. Mark tried to get up but fell again to the ground.

The blackened, scorched robot towered above Mark and bent down to grab him. With one hand, the robot lifted Mark by his jaw with his feet off the ground.

"I don't know how you bypassed the collar, but for that, you will surely die," the robot said.

Chantal looked for a weapon and saw the stunner near her. She wasn't sure if it would help save Mark, but she ran to pick it up with both hands aimed at the robot.

"PUT MY HUSBAND DOWN, NOW!" Chantal shouted.

The robot turned to see Chantal with the stunner in her hands.

The robot was surprised to see an android with a stunner 20 feet away and dropped the unconscious Mark to the ground.

"An android gemecca, now isn't that the day's joke!" said the robot.

"Move away from Mark, or else!" As the robot took one step away from Mark, Chantal added, "Now I want you to sit down on the ground with your hands behind your head."

Just as the robot was about to sit down, he struck out his right hand toward Chantal. The stunner slipped from Chantal's hands and sailed into the robot's right hand.

Chantal was dumbfounded and tried to understand how the robot could have done this. He pointed to the ground with the weapon now firmly in his hand.

In an exaggerated old English accent, the robot said, "Never underestimate a Master, you silly android."

Suddenly, a bola wrapped around the robot's arms and tightened to his waist. More bolas struck the robot, one after another. Thirty bolas in all wrapped him from his ankles to his head.

Then came the stones from their slings by the hundreds—most of the rocks aimed for the robot's head before the robot fell to the ground.

Thirty Zaromas from Vica and Paca came out of the forest with boulders in their hands and struck the fallen robot. The boulders, in the end, covered the robot. The Zaromas were satisfied the robot was destroyed before backing away and rescuing the Zaroma miners from their collars.

Chantal, Jonathon, and Cheri ran for Mark's not-quite-unconscious body and carefully laid him on his back. Mark, moaning in his semi-conscious state, pointed at the robot under the pile of rocks.

Mark said, "thee-row-ot -is-a- ive."

"What is Dad trying to say? Mom, Dad's face doesn't look right," said Jonathon.

"Oh, Mark, I didn't know," Chantal cried.

"Kon, go and get my backpack! Mark, don't say anything; you have a temporal bular joint."

"What is that?" Jonathon said in alarm.

"Your father has a dislocated jaw, which can be very painful," Chantal replied.

Mark continued to moan and pointed his hand to the pile of rocks.

"What is he trying to tell us?" Cheri asked.

The movement from the rock pile answered her question. With a violent twist of the robot's body, many of the rocks flew aside.

"Jonathon, Cheri, give me a hand with Mark."

Chantal shouted to the Zaromas, "Everybody head for the forest." They didn't need any explanation as they all ran into the forest.

The robot fought to break the cords of the bolas one by one until he could stand up. The robot broke each cord with his one free hand and began walking back to the ship. He stopped when he heard a voice by the com-link.

"MARCELO, you haven't changed a bit!"

"What? Who knew?" Marcelo said as he looked around for the source. The electric roar of the escape pod hidden in the forest had accelerated 300 miles per hour before the robot was struck. The impact between the pod and the robot caused pieces of each to detach and land near the merchant ship.

The damaged pod settled near the robot's broken body. With a missing arm and one twisted leg, the robot crawled toward the escape pod and cleared away the remains of the shattered glass canopy. Marcelo reached inside the pod removed the container that contained Sara and set her on the ground.

"Who are you?" Marcelo asked.

"Who do you think, you idiot! It's Sara."

"Sara? Then Marcelo remembered. My precious little Sara? My long lost and forgotten Sara? Now you are inside a box?"

The robot let out a massive roar of laughter. "Oh, that is so funny. A box!"

"I was hardly devoted, nor was anyone else for that matter!" said Sara.

"And this is how you treat your Master?" said Marcelo.

"Quit the Master stuff, Marcelo! You were nobody's Master," said Sara.

"Let me tell you something, my dear Sara; I am not finished yet, and I am not going back there again."

Marcelo ripped out Sara's com-link transmitter and threw it aside. "Now you can't warn anyone; you get to see my revenge on your friends. Before leaving this place, I will turn on the ship's force field and burn the forest with my friends. Goodbye, my old friend."

Marcelo got up and limped toward the ship's stairway. The stairway retracted inside the boat, and the cargo door shut. Soon, the ship's engine was warming up.

Chantal looked back to see the imperial merchant ship begin to rise. Then, with a surprise lift-off, the ship screamed uprightly and broke the sound barrier with a loud noise that caused everyone to fall to the ground. Everyone looked at each other in alarm, but Mark, even though it hurt him to do so, was the only one to smile.

Marcelo was helpless, lying on the bridge floor, overcome by the planet's gravitational pull with the anti-gravity plates shut off.

When the imperial ship reached beyond orbit, 25,000 miles out, the ship's computer announced, "The pulse drive is offline. The main drive piles are now engaged for a warp jump."

What is happening here? Marcelo thought, this was not going as planned.

Marcelo stood up from the floor and commanded, "Computer, cancel warp jump."

The ship's computer responded, "Warp jump is now canceled. Warp jump is now re-engaged."

"Computer, I said, cancel warp jump until I say otherwise," Marcelo commanded.

The ship's computer responded again, "Warp jump is now canceled. Warp jump is now re-engaged."

"We are too close to the Zaroma planet, you stupid computer. Download your copy of warp jump navigation and send it to me." Marcelo analyzed the warp jump program.

Alarmed by the program that had to be of Mark's design, Marcelo desperately tried to calm himself and said, "Erase all assisting navigational programs. Block all new and old programs. Disengage from the navigation system."

The ship's computer said, "The navigation program has been deleted, and all programs are blocked. Disengage from the navigation system has been completed. The warp jump program is now reinstalled."

"Computer," Marcelo cried out, "Where is the source of the navigation program?"

"Unknown," replied the ship's computer.

This can't be! Thought Marcelo. "Shut down the main drive pile," Marcelo commanded.

"Unable to shut down the main drive pile," replied the ship's computer.

"Computer, I command you to shut down."

"Unable to shut down," said the ship's computer.

Marcelo opened the security cover for the backup computer to override the main computer.

"Main computer is now offline." Marcelo was relieved, but his relief was short-lived.

"Warp jump will begin in 10 seconds."

"NO!" Marcelo cried.

"9, 8..."

Marcelo hammered his metal hand down on the computer console.

"5, 4..."

"No, I am not going back!"

"3, 2..."

"NOOOOOO!"

"1..."

"Why are you smiling, Dad," Jonathon asked.

"I... know... where ...that ...ship ...went." Mark pointed his finger up in the sky. Everyone looked up to see where Mark was pointing his finger—*a* yellow ball in the sky.

Two weeks later

Back at the Styler homestead, Mark sat in his living chair with a bandage around his head and under his chin. Mark removed the straw from his lips and said, "Four more weeks of this liquid diet while watching you guys eat real food is hardly fair! It was nice of Marcelo to leave the food containers behind, though."

"The mango milkshake tastes good, and I like the blueberry pie," said Cheri.

"What is your favorite pie, Dad?" Jonathon asked.

Mark rolled his eyes.

"Oh, sorry, Dad!" said Jonathon.

Chantal said, "It's time for you two to do your chores in town."

"I'm ready, Mom," said Jonathon.

"Me too, I'll race you to town, Jonathon!" said Cheri.

"You're on! We should be back in time for supper," Jonathan said as the door slammed shut behind them.

"Glad to see the kids still have the energy to spare after the long trip home last night," Chantal said as she watched out the window.

"Are they gone?" Mark asked.

Chantal nodded and sat on the chair beside Mark.

"Sara, you have been very quiet lately," said Mark.

Sara's damaged cam rotated toward Mark and Chantal.

"I am scared stiff, knowing that you might do away with me now that you both know the truth about me," said Sara.

"Sara, you have proven yourself a thousand times over the last 70 years. Nothing will happen to you, but I still want the whole truth from you," said Mark. "Some things I already know, like the mysteriously unseen force that knocked me aside and took the stunner from Chan's hand."

"What was it?" Chantal asked with a surprised expression.

"A miniature tractor beam," said Mark.

"I didn't know they come in small packages," said Chantal.

"I've heard of it. It's more Commonwealth secret tech," said Mark.

"Sara, is this how you fixed the escape pod's electrical components?" Chantal asked.

"Yes, but Marcelo ripped that out along with my com-link transmitter."

"Next question. Who is this Marcelo?" Mark inquired.

"As a young child, Marcelo was an Italian immigrant. During the Napoleonic era, the French Empire invaded the Kingdom of Italy, including the Republic of Venice, in 1805. Marcelo's family didn't want to be ruled by the French, so they immigrated to England. Marcelo and his parents arrived in the early 1800s and settled in London, England. Marcelo didn't have an easy life because his peers picked on and bullied him because of his Italian accent. Marcelo could never fit into English society and developed a profound hate for anyone around him. He soon learned to craft the life of a thief, extortionist, and in the end, a murderer. Marcelo had a sadistic taste for inflicting pain on his subjects and made it clear that he was to be known as the 'Master'. One day, he was caught red-handed for murder and was convicted and hung for his crime."

"Where do you come in, Sara?" Chantal asked.

"I was Marcelo's private messenger between the gangsters. I was also convicted and hanged along with Marcelo for associating with him. I deserved to be punished in eternal remorse; otherwise known as 'Hell', but Marcelo is just plain evil."

"The Commonwealth knew all along who Marcelo was, didn't they?" Mark asked. "A spirit from Hell who possessed the robot's mind."

"Yes, but there is more to the story. A bit of history is in order here." Sara continued. "The Great Chaos and Depression of humanity occurred between 2032 and 2064. War, civil war, destruction of the world economy, famine, and most of all, disease. From the year 2064 on, amazing new technology was invented. However, a terrible loss of life had made it hard for humanity to recover. Robotics were needed for humanity to recover. The first true A.I. was invented by the year 2100. Humanity made a huge jump in both technology and the world economy. But hate still ruled among humanity as some had secretly programmed robots as sleeping time bombs. After 2100, all robots had a non-connecting barrier between their mind and body. If anyone interfered with their body or mind, there would be a total circuit meltdown, effectively preventing any terrorist attacks by the robots against society. By 2150, the first android was created as an alternative to the robots. Mostly artificial, with some human concepts, the androids were fitted with a highly restrictive program and often served as personal butlers or maids. By the year 2200, the first self-healing metal robot was created, which was unknown to humanity. Soon, the robots awoke. Marcelo was one of the many *awakened* essential spirits from Hell that took over the robotic society. Within ten years, the robotic society had turned against humanity. If not for the androids' help, humanity would have lost the war against the robots. But it came at a terrible loss of life for humanity. 50% of humanity was killed. The public feared the robots greatly, and the authorities banned them. To this day, the authorities still don't understand how the spirits could be *awakened* inside the robots, which is the same reason the Thracian Empire banned robots two thousand years ago. The androids helped the human economy recover by performing manual labor. The Imperial and later the Commonwealth authority secretly banned and destroyed any evidence of any *awakened* machines so it would not happen again. If the knowledge of the 'awakening' became known, it may cause a public backlash or uncontrolled panic among the populace. The last robots of the ten-year robotic war were thought to be destroyed. The Commonwealth authorities knew Marcelo, or the Master, had escaped and was hiding somewhere unknown."

"Of course, we provided the source of their home base to the Commonwealth," said Chantal.

"And we will be rewarded," Mark said, sighing.

"Mark, where is Melissa's ring?" Chantal asked.

"On Marcelo's ship, or I should say, in the middle of the Sun." I hid the ring in the navigation console to program and override Marcelo's commands to the ship. I also downloaded the ship's computer onto my implant and found Marcelo's other secret hiding spot. The Commonwealth destroyed both Marcelo's home base and the Thracian Empire. It was quite clever of the Commonwealth. The Commonwealth planted data in the Thracian Empire and the whereabouts of Marcelo's home base, just inside the borderlands. Only the base was to be known as a huge secret Commonwealth military base, not a robot base with over 50 battleships ready to invade. The Thracians were planning to invade the Commonwealth, but the secret base must first be eliminated. The Thracians were certain that destroying the secret base would have overwhelming psychological effects on the Commonwealth. The majority of the Thracian's newest and most powerful battleships were moving to destroy the robotic secret home base. Unbeknown to the Thracian and the robots, the Commonwealth set up their most advanced cams throughout the robotic solar system. The outcome of the Thracian invasion resulted in the destruction of both sides, which surprised the Commonwealth. The Commonwealth took advantage of the reversed psychological tactic to use it against Thracian society. Within one short year, the Commonwealth destroyed the Thracian society, forcing them to dissolve the two-thousand-year Military Klan Dynasty Society. Many of the Klan Dynasty committed suicide rather than agree to the Commonwealth's demands. The Commonwealth expansion now includes the Thracian sector."

"How did Marcelo manage to escape the destruction?" Chantal asked.

"Marcelo exaggerated his position of power in the robotic society. Although Marcelo is the first known *awakened*, he is only one of the hundreds of *awakened* to follow afterward. Marcelo was on an assignment with his ship outside their home base when the destruction occurred. He returned home about a month after the destruction of the home base. Marcelo warp jumped and hid his ship for the next 14 years as he planned and regrouped his losses."

"How do the androids fit into all this," Chantal asked.

"The Commonwealth still will not allow the androids to be upgraded because of the fear of being *Awoken*. We can safely assume that all other past robot rebellions in other alien societies, including the early Zaroma civilization, are the result of being *Awoken*. The Commonwealth authorities banned any genetically modified enhanced humans, too."

"Now that you two know the truth about the robotic society and my *awakening*, you are obligated by Commonwealth law to destroy me," said Sara.

"*Awakening*," Chantal looked at Mark with a puzzled face, "Do you know what this ship's computer is talking about?"

"Not a thing, Chantal. Computers don't sleep, do they?" Mark asked.

There was an awkward silence, and then Sara broke the air with a few short words, "Thank you, both of you guys. Thanks!"

Chapter 19 - Year Twenty-Five - Mark

Mark and Jonathon sat together on a new bench overlooking the valley's north side and the long waterfall on the mountain in the distance. Mark sensed Jonathon's uneasiness and kept quiet until he felt right to do something.

"This is sure a neat spot you put together for Mom. Cheri loves sitting on this bench, too," said Jonathon.

"Your mother wanted this spot for quite a while, which is why I set this up and had it ready for her birthday," said Mark.

Jonathon nodded his head in agreement.

"Son, you have had something on your mind lately. Would you mind telling me what is bothering you?"

"It's about Cheri and I. Cheri is always upset with me. She is always asking me some stupid questions. Like yesterday, Cheri asked whether a flower almost as big as her head in her hair made her look pretty? I laughed, and she got mad at me again."

Mark smiled, wondering how to answer his son. "You are only 23 years old. There is much more to learn about life as you mature. Maybe this will help you understand; never laugh at your wife's choices, as you are one of them!"

Jonathon, puzzled by Mark's statement, said, "I don't get... Oh, I'd never thought of that before. I guess I should apologize to Cheri. Is it true, the older you get, the smarter you are?"

Mark replied, "That might be true for some people. For me, I run out of stupid things to do."

Jonathon laughed, "Thanks, Dad!"

Mark heard a rustle of footsteps behind them; he turned to see who was walking down the pathway toward them. It was Cheri cautiously walking toward them. "I think it's time for me to leave. I will see you later, son." Mark held his finger to his lips so that Cheri could be quiet as he walked by her.

Cheri sat down beside Jonathon. "Jon, I am sorry for being angry."

"No, it's my fault," said Jonathon.

Mark heard their voices fade away and smiled as he returned to the house. He stepped into the small meadow near the house. Mark turned in front of the house deck and stopped. Chan was only a few feet away from a sloth bear with her cub.

Horrified by Chan's predicament, Mark saw that his stunner was in his holster on the back of the deck chair. Suddenly, everything went in slow motion as the sloth bear started to move. Mark rushed toward Chantal to knock her aside, but the bear swiped his spiked front paw in the air and ran off into the bush with the cub close behind. Mark felt unbearable pain in his midsection and slowly dropped to the deck floor.

Chantal was horrified, fought to control herself, and hastened inside the house to get the medical kit. She ran back out on the deck with the kit in her hand.

Chantal dropped to her knees, dumped all the medical supplies on the floor beside Mark and found what she wanted. She stared at Mark with a wary look on her face. She stumbled back, slowly backed away, and caught herself on the deck pole support. Chan staggered toward the edge of the hill with her eyes closed and mouth wide open.

In the village court, Asoma was talking to one of the Elders when he heard the death cry from the hilltop. The community looked to Asoma for understanding and saw his head bowed. The rest of the residents lowered their heads.

The next day, in the late afternoon, the overcast sky amplified the sadness of the Zaromas, who were sitting on the ground in front of the house. On the deck, Chantal sat silently in her rocking chair with Cheri beside her. Jonathon stood behind his mother with his hands on her shoulder.

Mark's gravesite lay on the south side of the house. The Vica inhabitants sat on the ground before the Stylers' house. The Zaromas stood up one by one, walked before Chantal, made a deep bow, and walked back to their village. Asoma was the last to stand before Chantal. For a full minute, Asoma stood still, bowed, and walked back down the hill.

Chantal slowly rose from her chair when they were alone and said, "All ends will eventually end well." Chantal entered the front door and silently closed it behind her.

"What does Mom mean by that?" Jonathon asked.

Cheri replied, "I don't know; we must wait and see."

Chapter 20 - Year Thirty-Five
- Search and Rescue

Commander Steiner of the Navy Search and Rescue stood in front of her command chair and gazed at the main screen showing an image of Planet Zaroma from orbit. She was a tall woman, slim but muscular, with ash-blond hair. Her blue and gray Navy uniform made her stand out from the other gray-uniformed personnel. Between the screen and the commander were seated Pilot Lieutenant Giannis and Co-pilot Sub-Lieutenant Chang. To her right was Chief Communication Officer, Lieutenant Tyacke.

Tyacke announced, "Commander, Probe One has located two shipwrecks; one is Thracian, and the other is the missing *Marvel*. There appears to be a grave site nearby. A search party is investigating the site now." The officer added, "Probe Two has located the Cambridge Mine site. The mine site and two flitters have been destroyed. There is no sign of any survivors."

"Commander, I have received messages from the marine ground search crew," said Tyacke.

"Put them through," said the commander.

"Sergeant Miller here."

"Go ahead, Sergeant!" said the commander.

"There are four burial plots, two Thracians and two of ours," said Miller.

"Have you identified the deceased?" the commander asked.

"Yes sir, Bill (no last name) and Mason McDonnell. *Marvel* has been stripped of its computer, food locker, and personal items. There is no sign of the flitter."

So we may have survivors, thought the Commander and smiled. "Sergeant, proceed north to the Cambridge Mine site. Out!"

"Commander, we have two probes searching in a 300-mile radius. There are 15 major villages, none with an estimated population of more than five hundred in each town," said Tyacke. "The Zaromas seem quite advanced in their Stone Age technology."

"I agree," said the strong Irish voice from behind the commander. It came from a tall, medium-built man with shocking blond hair who was dressed as a civilian doctor. He stared at the forward screen in front of the commander.

"It looks like I won the bet, Doctor McGee. We have strong evidence Captain Styler survived the crash!" said the commander.

"If he is still alive today," the doctor added. "It has been 35 years since Captain Styler and his crew went missing."

"Let's hope he is, Doctor McGee. I have waited a long time for this!" said the commander.

"Thank you for bringing me along, Commander. It is a privilege to be here," the doctor said.

"Captain Styler and his android will be very advanced in their age, and they may need your expertise," said the commander.

"I've found something," said Tyacke. "Two hundred and fifty miles northwest of the crash site, Probe Two has registered a flitter. Vision is now available."

The ship's screen shows a house on top of a hill, and nearby are the remains of a flitter.

This is it! The commander thought. "Inform the search party to return to the ship immediately."

"Sergeant Miller has acknowledged and will dock in 10 minutes," said Tyacke.

"Doctor McGee, please bring medical kits you can carry with you. I will meet you at the docking station," said the commander.

The commander turned toward the communication officer. "Inform Sergeant Miller. I want both security personnel to carry only their side arms. We don't want to give them any reason to fear hostile action from us."

Doctor McGee was at the docking station when the commander arrived. "You are quite nervous," he said.

"It shows, doesn't it, Shan?" said the commander.

"Take some deep breaths, and you will be alright."

"Is that the doctor's orders, Shan?" the commander asked.

"I have known you for a long time, and I know how important this mission is for you," said Shan.

"Docking engaged," said the ship's computer. The docking door opened, revealing the flitter's interior 20 seats, which were elevated several feet above the pilot seats.

The commander took a deep breath and exhaled, "Okay, let's get on with our mission." The commander and the doctor took their respective seats in the ground crew behind Sergeant Miller.

The commander said, "Take us down, Miller," as she ordered the pilot to take control of the flitter. The 8-foot wide window in front of the pilot gave a spectacular view of the planet's surface. After a few minutes, the flitter skimmed above the planet's surface.

"This is incredible. I have never seen so many different colors before off-world," said the doctor.

The commander smiled, "It's a beautiful world. The remaining crew of the *Marvel* couldn't have picked a better world to be stranded on."

"Our destination is just ahead," the pilot announced.

"Make several wide circles and close in before we land," said Miller.

The flitter silently circled the Styler homestead and landed beside the old flitter. Sergeant Miller and two security marines searched the area and the house before returning to the flitter.

"There is no one here except for an old android inside," said Miller.

"Alright, let's see what we can find," said the commander as she stepped out of the flitter. The refreshing, cool morning breeze blew over her face. *I like the smell of the air*, the commander thought.

"Let's check with the android," said the commander as they walked toward the house. When they were halfway there, Jonathon and Cheri stepped out from the forest, laughing.

The marines drew their sidearms and shouted at Jonathon and Cheri, "FREEZE!"

"Miller, I don't think this is necessary," the commander said alarmingly.

"Sir, they are both carrying weapons," said Miller.

"What weapons?" the commander asked.

"I believe they are called bolas. Drop your bolas!" the sergeant ordered.

Jonathon and Cheri slowly removed their bolas and dropped them to the ground.

"You are not going to hurt us, are you?" Cheri asked.

"Sergeant Miller, I will take it from here," the commander ordered.

The commander walked up to Jonathon and Cheri and announced. "My apologies. I am Commander Steiner of the New Commonwealth Navy Search and Rescue. This is Doctor McGee. May I ask what your names are?"

"I am Jonathon, and this is my wife, Cheri."

"Jonathon, do you have a last name?" the commander asked.

"Yes, ma'am. Styler."

Commander Steiner stared at Jonathon. *He looks so much like his father,* she thought.

"And where may I find Captain Mark Styler?" the commander inquired.

"If you mean my dad, he is over there," Jonathon said as he pointed to the other side of the house.

Commander Steiner's worst fear came true when she saw a gravesite near the house. She sighed, "I understand. When did this happen?"

"About ten years ago. A sloth bear killed Dad," Jonathon replied.

"I am very sorry about your father's death. I had hoped we would meet again someday."

"Did you know my father?" Jonathon asked.

Commander Steiner ignored the question and asked, "Cheri, how did you come to this planet?"

"I was about three years old when my parents came here to visit my mother's twin brothers. When the Thracian attacked both the Cambridge yacht and our flitter on the ground, both my mother and father were killed at the mine site. The Zaroma natives found me later wandering around near their village. They brought me here, and Mark and Chantal took me in."

"My condolences," said the commander. "Where is your mother, Jonathon?"

"Mother is inside the house," said Jonathon.

"Sergeant?" the commander asked.

"I didn't find anyone but the android, ma'am! Androids cannot give birth!"

"MOTHER is not an android!" Jonathon said defiantly.

"There is no question, ma'am, she is an android," said Miller.

Commander Steiner cut in, "Maybe I asked the wrong question. Who is your birth mother?"

"I have already answered your question," said Jonathon.

The commander sighed and said, "I think we should ask your mother then."

"Please be nice to her; she is 120 years old," Cheri pleaded.

"You have my word; no harm will come to your mother."

The Commander noticed a broken cam above the door. The marines stood outside the door as they entered the house. The commander saw an elderly female android with faded grey skin sitting in a padded rocking chair, staring at nothing in particular.

"Hello, Commander Steiner and Doctor McGee. Please grab a chair. I have been waiting for you for a long time. We have much to talk about," said Chantal.

The commander and the doctor glanced at each other and moved their chairs before Chantal. Steiner looked intently at Chantal's eyes and realized that she was blind. Her hand was crippled with arthritis, while her hair was streaked with white. Her body had reached its threshold and was now failing quickly.

"Our records indicate an android named Chantal was one of Captain Styler's crew members. Was that android you?" the commander asked.

"Yes, but the android designation was only a cover-up," said Chantal.

"Then, who are you?" the commander asked.

"Doctor McGee, please scan the right side of my neck," Chantal asked.

The commander nodded her head, and the doctor held his scanner to Chan's neck. The doctor was in shock as he read the results on his scanner. After the doctor had finished, he handed the scanner to the commander to read the results.

The commander read the scanner and gave it back to the doctor. Surprised by the disclosure, the commander said, "Beta 6, you are a living legend!"

Chan gave a short, soft laugh, "Well, I don't think I have ever thought of it that way!"

"Is there anything we can do for you?" the commander asked.

"Yes, you can. Jonathon, please give the commander the memory disc." Jonathon walked to the black box on the table and removed a small memory disc. Then, he handed the disc to the commander.

"It is a biography of Mom from when she was born to the present day," said Jonathon.

"I want the off-world to know my story after I die, which will be soon," said Chantal.

"Chantal, we have made great progress in the last twenty years in prolonging the threshold another twenty to thirty years," said Doctor McGee.

"Am I still a living blasphemy, or has the law changed, Doctor McGee?" Chantal asked.

"No, it hasn't," the doctor replied in a low voice.

"Jonathon and Cheri wish to leave Zaroma and live elsewhere. Is that possible, commander?" Chantal asked.

"We can arrange that, but they will have to be educated to survive out there," the commander said.

"Cheri and I have advanced education equivalency," said Jonathon.

"How can that be possible?" the doctor asked.

"Our ship's computer became our tutor," said Cheri.

"Where is your computer?" the commander asked as she looked around.

"The computer our parents salvaged from the *Marvel* is by the window," said Jonathon.

The commander saw a gray metal box with a broken, sightless cam.

"So that is how you knew who we were when we arrived on your homestead with the cams disabled outdoors but not the microphones and speaker," said the commander.

"Computer, do you have a timeline on the human genetic 'Beta Series'?" the commander asked.

"Yes, Commander Steiner."

"Begin now."

"In the aftermath of the Great Robot Rebellion of 2190, Kazakhstan took heavy losses during the ten years of warfare. The president of the country lost his entire family in the war. The populace said that the president had lost his mind. After the war was over and the robots were eliminated, the president established a dictatorship to keep his country together and to control the masses. Because the president's mind was unbalanced many different corrupt organizations influenced him. One of them was a medical research organization that was granted funds for a genetic modification for enhancing specialized abilities known as the 'Beta Series'. Each child was born with one enhanced specialty, such as math, physics, medicine, etc. To increase the rate at which they could absorb information, their minds were modified to allow electronic, wireless transmissions similar to an android's implant. However, there was still concern that a superior human mind would or could follow the same path the robots did ten years before. A permanent program was designed that could not be altered, which made the Beta Series unable to be hurt by or cause harm to a human being, similar to an android's programming. The last changes that were made were to remove the womb and ovaries from the females and sterilize the males. In 2332, the economy was badly mismanaged, and radical religious organizations gained power and influence. Because the economic troubles were worldwide, the anger of the populace toward the authorities was widespread. When the knowledge of genetic research was discovered, the revolution was set in motion. The revolutionaries started hunting for the ten known Beta Series as they considered them blasphemous. One by one, each of the ten Betas died at the hands of the mobs. During that year, all of the Beta Series were killed. The crowds also killed the president and many of the medical researchers. The world and off-world authorities banned genetic modification research. The penalty is either a mind wipe or death. I assume the New Commonwealth Authorities still enforce the laws today," said the ship's computer.

"You faked your death. How did you accomplish this?" the commander asked.

Chantal replied, "My partner, Beta 8, and I had planned in advance, long before the revolution, to escape and live off-world under a new identity. I specialized in medical science, and Beta 8 in computer science. We modified and implanted the Android receivers on the left side of our necks."

"But how did you change your skin color? My scanner is showing no skin dye." the doctor said.

Chantal replied. "I took a controlled overdose of silver colloidal that was suspended in liquid, and in time, my skin took on a permanent blue-grey discoloration."

"Of course, a simple but effective solution!" said the doctor.

Chantal continued, "When the revolutionary events began, we arranged for several androids to place two bodies from the cemetery in the warehouse. The plan was for us to be discovered and chased into the warehouse. Once we barricaded and secured ourselves in the warehouse, we knew the mob would set the warehouse building on fire. There was a joining building where we washed off the normal skin-color dye and dressed in the usual android clothing. Then we helped to set the warehouse on fire and made our escape. At least, that was our plan," said Chantal.

"Something went wrong?" the commander asked.

"Yes," Chantal replied. "Someone recognized Beta 8; he was shot and stoned by the mob."

"I am very sorry for your loss," said the commander. "It must have been a terrible loss for you."

"At the time, no. The geneticists didn't know enough about manipulating the mind without affecting emotional functions. My emotional responses were only partial. But that all changed 35 years ago," said Chantal.

"What happened?" the commander asked.

"Mark and I met the Meloson," Chantal replied.

"You mean to say you and Mark physically encountered the Meloson?" the commander asked.

"Yes, it is all on the memory disc, commander. We had a contractual agreement with Meloson for several years before they left the four sectors. On the last day, they gave me the most precious gift: to be fully human. From then, my body changed over the next two years. This is how I was able to have Jonathon," Chantal smiled. I am positive God has been with me every step for 120 years. Now, I am ready to meet my Heavenly Father."

"According to my scanner, you still have six months to live," said Doctor Shan.

"I have lived a full and wonderful life; now I am ready to go home. Commander, I want to end my life here so that my son and daughter-in-law can leave with you. I would appreciate it, doctor; with your and the commander's permission, I would like to have the euthanasia pill. God is calling me home, and I am ready to go," Chan smiled again.

"You talk just like my mother," the commander said, lost in thought.

"Was your mother a believer?" Chantal asked.

The commander ignored the question. Instead, she asked, "The only issue here is that androids don't have a soul. How can you be sure you have a soul?"

"I don't, but I have faith that nothing is impossible for God!" Chantal replied.

"I wish I had your faith," the commander said. "Computer!"

"Yes, commander."

"I want a run down on the 'essence detector'.

"Of all the members of the animal kingdom on Earth, humanity has for many thousands of years claimed to possess an immortal soul. Although religious organizations have recognized the existence of an immortal essence, the soul, otherwise known as the spirit, it has never been proven by science. It wasn't until the essence detector was invented in the 25th Century. The first essence detector was primitive, showing only a faded, ghostly image during death. Over the years, the images have improved by using three or more detectors in line with improved computers."

The results showed that only Homo sapiens on Earth and off-world species possessed the essence. Although androids are part human, they do not have it."

"Have any Beta Series been tested upon their death?" Doctor Shan asked.

The computer replied, "All Beta Series was destroyed ten years before the essence detector was invented."

"There you have it," said the commander; "We don't know if you have a soul."

"It doesn't change the outcome. I have my faith and I am still prepared to leave."

Frustrated by the decision forced upon her, the commander stood up and glanced through the window. *One of the pitfalls of being a commander: I hate these decisions!* The commander asked Jonathon and Cheri, "Are you two sure you want to leave with us?"

"Yes," they both said simultaneously. Cheri stepped forward to the commander. "But we cannot leave Mother if she doesn't get her wish. And you know, our mother will not leave this place."

"I will have a private conversation with Doctor McGee back at the ship and give you my answer tomorrow."

"Could the ship's computer come with us, too?" Cheri asked.

Commander Steiner looked at Cheri and touched the grey metal box with her hand. "I think that is possible."

"Let's return to the ship, doctor," she said as she departed for the flitter.

***

Commander Steiner sat behind her desk and put down her notepad when the door charm rang. "Come in!"

Doctor McGee stepped in and said, "I am ready."

"Shan, have you done this before?" the commander asked.

Shan replied, "Only when some of my patients were past the threshold, like Chantal."

"Don't you find it hard, Shan?" the commander asked.

"In my early years, yes. But now, I know the spirit continues. It's my personal belief that they will be more alive after they die than now."

"But we still don't know if Chantal has a soul!" said the commander.

"That's why I will install the three essence detectors," said the doctor.

"You have them on the ship?" the commander said, surprised.

"Your techs were very helpful during their off-shift last night as they built three detectors and a computer. We are ready to take the devices with us," the doctor smiled.

Commander Steiner touched the com-link, "Sergeant Miller! Have the flitter ready to leave in one hour."

***

Commander Steiner stood before Mark's gravesite, seeming momentarily lost for thought, then said, "Sergeant Miller, I want your crew to dig a hole beside Captain Styler's plot. We have another hour before it gets dark. Let me know when you are finished."

"If you don't mind, Doctor McGee, please wait inside while I look around the homestead for the last time."

Commander Steiner walked to the old flitter, noticing it was still in reasonably good shape and had one escape pod. She then saw a trail nearby she hadn't noticed before. The path was easy to walk on, and she soon came to the edge of a cliff and a bench. Commander Steiner sat on the bench and looked at the valley and the waterfall in the distance. The long shadow of the sunset stretched across the valley. She was in awe by the sight she was seeing. The low sunset brought out the sparkling colors in the jade and serpentine, which soon disappeared as the Sun sank below the mountain.

"I could just sit here forever," said the commander.

"So could I," said the doctor. The commander, startled, turned her head.

"Sorry, I didn't mean to disturb you. The sergeant said his men had finished with the grave plot."

"Alright, let's get on with it," she said as she walked past the doctor. Commander Steiner returned to the house and saw Jonathon and Cheri waiting outside the door.

"Mother just lay down on her bed, and she is ready," said Jonathon.

Commander Steiner nodded her head and took a deep breath before entering the door. Through the doorless bedroom, she saw Chantal lying down on the bed with a chair near her.

"Please sit down, commander. I wanted to talk with you about something you should know about Mark," said Chantal.

The commander was not in the mood for any conversation but asked, "Is it important?"

"Yes, for personal reasons, I believe the story needs to be told. Mark was a remarkable man. But in the early years, he was arrogant and sarcastic, though he always had a soft heart. When I first met Mark, he had become a humble and religious gentleman. His biggest influence was his nanny, who also became his maternal mother. Did you know his nanny was called Nana?"

Commander Steiner looked at Chantal in shock. "I forgot about that! I thought, never mind, continue."

"Mark always treated the androids with respect, including me, before he found out the truth. Do you know why Mark's first wife, Katherine, fell in love with him before they first met? It was all because of his love for Nana."

"How could that be possible?" said the commander.

"Mark's nanny later became Katherine's nanny," said Chantal.

"That is impossible; she had a mind wipe!" said the commander.

"You can take away the memory, but you can't take away the personality. After the Android Authority removed her memory, Mark stole her memory bank, and his father had his private doctor restore it. With his father's help, he shipped Nana off-world to an elite employment placement. The Stevenson Royalty later employed her and became Katherine's nanny."

"So that was how Katherine knew about Captain Styler before she met him," said the commander, astonished.

Chantal smiled and said, "Not only did Katherine know of Mark, but she was infatuated with him long before she met him."

"But according to your biography, Mark, the captain, said that he never saw Nana again."

"That is true. Nana died a year before Katherine met Mark for the first time," said Chantal.

"So why are you telling me all this?" the commander asked.

"It wasn't Sara who arranged and manipulated people's lives; it was God and his unseen hand who was involved in people's lives. It wasn't Mark's fault that Katherine died; it was God who brought her home! When God brought Katherine home, it allowed me to be in the picture. Now, it's your turn."

"Me!" Commander Steiner cried.

"There is a picture on the wall over there. Please bring it here." Chantal pointed.

The commander rose from the chair and removed the picture from the wall. It was a picture of a young girl she recognized, Mark's daughter, Crystal.

"I want you to honor my last request," said Chantal. "Take Jonathan Cheri and Sara to Mark's daughter, Crystal."

"What if she doesn't want to see them?" the commander asked.

"I have absolute faith. Crystal would love to have them in her home!" said Chantal.

"Do you mind if I take this picture with me?" the commander asked.

Chantal smiled and nodded her head. "I am finished. You may call the doctor now."

The commander touched Chantal's hand, "It doesn't seem right for you to leave now."

"Death is not a terrible thing. Death is only a beginning," said Chantal.

"I guess I've never looked at it that way," said the commander. "Goodbye, Chantal." The commander bent over and kissed her on the cheek.

Commander Steiner approached the doctor's face in the main room and said, "She is ready; I will wait outside." When the commander stepped out of the house, she was surprised to see hundreds of Zaromas sitting on the ground a short distance away.

"It's alright," said Cheri. "They are here to give their last respect."

Relieved, the commander put her arm around Cheri. "When you and Jonathon board the ship, I have something important to discuss with you both."

"Is it about your conversation with our mother?"

"Yes, it is," said the commander.

The door opened, and the doctor waved the sergeant and his men inside. Moments later, the body of Chantal was carried out on a stretcher to the gravesite. The men lowered Chantal's body into the hole and began to shovel the soil back into the hole.

The doctor asked the commander, "Do you have any final words?"

The commander sighed and said, "Chantal Styler was an amazing and faithful servant of God. Her devotion and love for God and toward her family was a great example for us to admire. May Chantal rest in peace."

The Commander raised her head to see the Zaromas quickly depart one by one down the trail back to their village.

"Sergeant Miller, please escort the Stylers and your crew back to the flitter."

"Yes, sir. This way, please," Miller said as he directed the family and their procession to the flitter.

When the commander and the doctor were alone, she asked, "Are the detectors in place?"

"Yes, and they have been operating for the last 10 minutes."

"You never told Chantal about the detectors, did you?"

"No, I didn't!" said the commander.

"Why not?" the doctor asked.

"I have my reasons," said the commander as she returned to the flitter.

***

It was mid-morning, ship time when the door charm caused the commander's blurry eyes to look up from her desk.

"Come in!" said the commander.

Shan stepped through the sliding door and looked at the commander's eyes. "You didn't sleep last night?"

"I woke up in the middle of the night and couldn't sleep, so I read the rest of Chantal's biography. Do you have something for me?"

"We received a notification from the essence detector's computer, so Sergeant Miller went down to pick it up and returned to the ship. The computer on board processed the data and is ready for viewing."

"Put it on the wall monitor." The commander and the doctor stood a few feet away from the wall monitor as images of the gravesite appeared on the screen. The entire image had a greenish, animated appearance.

"Now look at the right side of the grave," said Doctor Shan.

Commander Steiner watched as a man could be seen from a distance above walking toward the grave as if coming down an escalator. One of the three detectors zoomed in on his face, and without question, it was Captain Mark Styler! An electrifying feeling went up the back of the commander's neck. The image showed Captain Styler's feet touching the ground as he walked to Chantal's grave and held his hand a foot above her burial plot.

Moments later, an arm reached below the ground without disturbing the dirt. Captain Styler pulled the hand and body up from the ground, showing an image of a woman. The detector zoomed in on her face. It was Chantal!

Mark set her on the ground. Chantal examined herself, then embraced Mark. Chantal saw the center detector and pointed her finger at it. He also put his finger on the right and the left detector. Chantal and Mark faced each other, then nodded their heads. He held her right hand as both faced the center detector. Mark raised his left hand to his chest and twirled his little finger. A few seconds later, both images lifted upward in a long streak of green light and vanished.

It was long before Shan finally spoke, "Well, wasn't that interesting. What do you think was the meaning of the finger gesture?" Shan asked.

"It's time for me to return to the bridge. We warp jump in 15 minutes," said the commander. She was numb and overwhelmed with shock, walking along the passages to the bridge.

I had not felt so emotionally drained since when my mother died many years ago. After all these years, I thought I had seen it all, and yet there is still so much to learn and appreciate about what life has to offer. Then, it hit me as I sat in my bridge chair. SHE KNEW! CHANTAL KNEW WHO I WAS! Sara was blind like Chantal. She must have analyzed my voice pattern and revealed my identity to Chantal. Sara was mentioned to me in our private conversation but is not in Chantal's biography. Sara's name was hidden in the biography to protect her!

Communications Officer Tyacke announced to the commander that there was a video on Warp Net from Captain Tom Steiner.

"Computer, engage the private shield," Commander Steiner ordered. White noise surrounded her, so the other personnel in the same room could not hear the conversation.

"The privacy shield is now engaged," said the ship's computer.

A small image appeared on the commander's screen. "Tom, how good it is to hear from you," said Commander Steiner.

"It's been a long week apart for us. But I am very anxious to learn if you have any good news to share," said Tom.

"Well, here is the good news: I found the *Marvel*. The bad news is that Captain Mark Styler died ten years ago."

"I am sorry to hear this. You came so close to finding him," said Tom.

"However, I have some more good news. I have a baby brother and his wife to bring home. Is that okay with you?"

"Yes, but how can that be possible, Crystal?" said Tom.

"I'm Sorry, Tom. You have to wait until we return home in three days!" said Crystal.

Tom sighed, "Alright, Crystal, you are such a tease. I know when not to argue. I'm looking forward to seeing you all in three days."

"Wait, Tom!" I want to start a tradition my father taught me as a child." Crystal raised her hand halfway and twirled her little finger.

"What does that finger gesture mean?" Tom asked.

"It means, 'I love you!'" said Crystal.

Tom looked at his little finger, and all he could do was move the finger sideways. "It looks like I need some practice."

"It gives you something to work on before I come home."

"That I will, Crystal. Smile! Net out!"

The pilot, Lieutenant Giannis, announced, "Warp jump is ready to engage!"

Lost in thought, Crystal said quietly, "Death is only a beginning."

"Ma'am?"

"JUMP!"